Seismic and Design for Soil-Pile-Structure Interactions

Proceedings of a session sponsored by the
Committee on Geotechnical Earthquake Engineering of
The Geo-Institute of the
American Society of Civil Engineers
in conjunction with the ASCE National Convention in
Minneapolis, Minnesota, October 5-8, 1997

Edited by Shamsher Prakash

Geotechnical Special Publication No. 70

Published by the

ASCE *American Society of Civil Engineers*

1801 Alexander Bell Drive
Reston, VA 20191-4400

Abstract:

Pile foundations are used extensively to support structures in seismic regions around the world. For years, geotechnical engineers have neglected to consider the interaction soil-pile systems have with superstructures; while civil engineers have ignored the interaction of the foundation with the structure. In order to understand a structure's response to dynamic loads, the complete soil-pile-structure interaction must be analyzed. Thus, centrifuge models, shake table models, and analytical methods are used to explore the interactions among soil, piles, and structures.

Library of Congress Cataloging-in-Publication Data

Seismic analysis and design for soil-pile-structure interactions: proceedings of a session sponsored by the Committee on Geotechnical Earthquake Engineering of the Geo-Institute of the American Society of Civil Engineers in conjunction with the ASCE National Convention in Minneapolis, Minnesota, October 5-8, 1997 / edited by Shamsher Prakash.

p.   cm. -- (Geotechnical special publication ; no. 70)
Includes indexes.
ISBN 0-7844-0287-6
1. Soil-structure interaction-Congresses. 2. Piling (Civil engineering)--Congresses. 3. Earthquake resistant design--Congresses. I. Prakash, Shamsher. II. American Society of Civil Engineers. Committee on Geotechnical Earthquake Engineering. III. ASCE National Convention (1997 : Minneapolis, Minn.) IV. Series.
TA711.5.S45 1997   97-28739
624.1'762--dc21 CIP

GEOTECHNICAL SPECIAL PUBLICATIONS

1) TERZAGHI LECTURES
2) GEOTECHNICAL ASPECTS OF STIFF AND HARD CLAYS
3) LANDSLIDE DAMS: PROCESSES, RISK, AND MITIGATION
4) TIEBACKS FOR BULKHEADS
5) SETTLEMENT OF SHALLOW FOUNDATION ON COHESIONLESS SOILS: DESIGN AND PERFORMANCE
6) USE OF IN SITU TESTS IN GEOTECHNICAL ENGINEERING
7) TIMBER BULKHEADS
8) FOUNDATIONS FOR TRANSMISSION LINE TOWERS
9) FOUNDATIONS AND EXCAVATIONS IN DECOMPOSED ROCK OF THE PIEDMONT PROVINCE
10) ENGINEERING ASPECTS OF SOIL EROSION, DISPERSIVE CLAYS AND LOESS
11) DYNAMIC RESPONSE OF PILE FOUNDATIONS— EXPERIMENT, ANALYSIS AND OBSERVATION
12) SOIL IMPROVEMENT - A TEN YEAR UPDATE
13) GEOTECHNICAL PRACTICE FOR SOLID WASTE DISPOSAL '87
14) GEOTECHNICAL ASPECTS OF KARST TERRAINS
15) MEASURED PERFORMANCE SHALLOW FOUNDATIONS
16) SPECIAL TOPICS IN FOUNDATIONS
17) SOIL PROPERTIES EVALUATION FROM CENTRIFUGAL MODELS
18) GEOSYNTHETICS FOR SOIL IMPROVEMENT
19) MINE INDUCED SUBSIDENCE: EFFECTS ON ENGINEERED STRUCTURES
20) EARTHQUAKE ENGINEERING & SOIL DYNAMICS (II)
21) HYDRAULIC FILL STRUCTURES
22) FOUNDATION ENGINEERING
23) PREDICTED AND OBSERVED AXIAL BEHAVIOR OF PILES
24) RESILIENT MODULI OF SOILS: LABORATORY CONDITIONS
25) DESIGN AND PERFORMANCE OF EARTH RETAINING STRUCTURES
26) WASTE CONTAINMENT SYSTEMS: CONSTRUCTION, REGULATION, AND PERFORMANCE
27) GEOTECHNICAL ENGINEERING CONGRESS
28) DETECTION OF AND CONSTRUCTION AT THE SOIL/ROCK INTERFACE
29) RECENT ADVANCES IN INSTRUMENTATION, DATA ACQUISITION AND TESTING IN SOIL DYNAMICS
30) GROUTING, SOIL IMPROVEMENT AND GEOSYNTHETICS
31) STABILITY AND PERFORMANCE OF SLOPES AND EMBANKMENTS II (A 25-YEAR PERSPECTIVE)
32) EMBANKMENT DAMS-JAMES L. SHERARD CONTRIBUTIONS
33) EXCAVATION AND SUPPORT FOR THE URBAN INFRASTRUCTURE
34) PILES UNDER DYNAMIC LOADS
35) GEOTECHNICAL PRACTICE IN DAM REHABILITATION
36) FLY ASH FOR SOIL IMPROVEMENT
37) ADVANCES IN SITE CHARACTERIZATION: DATA ACQUISITION, DATA MANAGEMENT AND DATA INTERPRETATION

PREFACE

Pile foundations are used extensively to support buildings and other structures. Earthquakes may cause dynamic loads on such structures. The response of pile foundations to these dynamic loads is extremely complex. Soil behavior is non-linear during earthquakes. There has been interest in this subject throughout the world. In buildings and other structures, the interaction of superstructure becomes extremely important. Studies on piles during earthquakes is difficult because earthquakes cannot be made to order! Therefore, recourse is made to alternate studies, e.g. on centrifuge models and/or shake table models and analytical solutions. In many cases, analysis of piles is carried out by neglecting the superstructure effects by the geotechnical engineers and analysis of structures is performed considering them fixed at their base. In a realistic analysis, soil-pile-structure interactions need to be considered.

The objective of this session was to address this problem in terms of practice in analysis and design of pile foundations under dynamic loads and focus on the unsolved issues. The papers were, therefore, invited from authors both within and outside the USA. This session was held at the ASCE Fall Convention in Minneapolis, MN on October 6, 1997 and was sponsored by the Soil Dynamics Committee (now the Geotechnical Earthquake Engineering Committee) of the Geotechnical Engineering Division of ASCE (now The Geo-Institute of the ASCE).

It is the current practice of The Geo-Institute that each paper published in a Special Technical Publication (STP) be reviewed for its content and quality. These special technical publications are intended to reinforce the programs presented at convention sessions or specialty conferences and to contain papers that are timely and may be controversial to some extent. Because of the need to have the STP available at the convention, time available for reviews is generally not as long and reviews may not be as comprehensive as those given to papers submitted to the Journal of the Institute. These STP reviews ordinarily are carried out within a three month time frame. Therefore, it should be recognized that there is a difference in the purpose of contributions to the special technical publications as compared to those in the Journal. In accordance with ASCE policy, all papers published in this volume are eligible for discussion in the Journal of Geotechnical and Geoenvironmental Engineering and are eligible for ASCE awards. Reviews of papers published in this volume were conducted by the Geotechnical Earthquake Engineering Committee of The Geo-Institute. The following committee members or cooperating persons from the general membership reviewed these papers:

John Charles   Panos Dakoulas
David Frost   Phil Gould
Mary Ellen Hynes   T. Kagawa
Nozar Kishi   Sanjeev Kumar
T. Nogami   V.K. Puri
Jay Shen

Personal thanks go to Panos Dakoulas, Chairman of the Geotechnical Earthquake Engineering Committee, for his help and support. I want to thank the body of experts who gave both the time and effort in reviewing the papers. Last but not least, thanks are due to all the authors who kindly accepted the invitation to contribute to this volume and to the session in Minneapolis and the participants in the session and their discussions.

Shamsher Prakash, F. ASCE
Professor of Civil Engineering
University of Missouri-Rolla
Rolla, Missouri

Session Organizer and Editor

TABLE OF CONTENTS

Soil-Pile-Structure Interactions

W.D. Liam Finn[1], G. Wu[2] and T. Thavaraj[3]

Abstract

A computationally efficient method is presented for the 3-D analysis of pile foundations under strong earthquake shaking. The key elements of the method are: efficient nonlinear 3-D analysis using a reduced 3-D equation to describe the half-space enclosing the piles, direct analysis of the pile group including kinematic and inertial interaction, and the ability to establish time-dependent stiffness and damping factors during earthquake shaking. The approach provides a comprehensive understanding of how pile foundations behave during strong seismic shaking, when non-linear soil behaviour is significant. The proposed method has been validated against existing elastic solutions for linear response and against nonlinear response of single piles and pile groups in centrifuge tests under strong shaking against centrifuge test data for single piles and pile groups for strong shaking.

Introduction

Seismic soil-structure interaction analysis involving pile foundations is one of the more complex problems in geotechnical earthquake engineering. A very common example is the 3-D analysis of a pile foundation for a bridge abutment. The analysis involves modelling soil-pile-soil interaction, the effects of the pile cap, nonlinear soil response, and in many cases seismically induced porewater pressures. There are many approaches to solving the dynamic response of pile foundations. Novak (1991) presented an extensive critical review of the more widely accepted

[1]Professor, Department of Civil Engineering, University of British Columbia, 2324 Main Mall, Vancouver, B.C. Canada
[2]Agra Earth & Environmental Ltd., 2227 Douglas Road, Burnaby, B.C. Canada
[3]Graduate Student, Dept. of Civil Engineering, University of British Columbia, 2324 Main Mall, Vancouver, B.C. Canada

methods of analysis. The dynamic characteristics of piles in a group may be quite different from the static characteristics, because the dynamic interaction factors of pile groups are strongly frequency-dependent. Therefore, pile group response cannot be deduced from single pile response without taking dynamic pile-soil-pile interaction into account.

The methods used in practice for direct seismic analysis of pile groups are based on linear elastic behaviour. Complete elastic analyses have been conducted using 3-D boundary element formulations (El-Marsafawi et al., 1992a, 1992b), but they require substantial computing time. They are exact for elastic isotropic conditions. However, they cannot take into account the nonlinear behaviour of soil under strong shaking or the effect of seismically induced porewater pressures on dynamic response. The reduction in soil stiffness and the increase in damping associated with strong shaking are sometimes modelled crudely in these analyses by making arbitrary reductions in the shear moduli and arbitrarily increasing the viscous damping. For this reason, the results of these studies have not proved very useful for the response of pile foundations to earthquake loading.

To simplify group analysis, El-Marsafawi et al. (1992a, 1992b) developed approximate procedures for estimating elastic dynamic interaction factors based on boundary element analysis. They extended the studies of Kaynia and Kausel (1982), Davies et al. (1985), and Gazetas (1991a, 1991b), using the general 3-D formulation developed by Kaynia and Kausel (1982).

The offshore industry pioneered the seismic design of pile foundations by estimating the nonlinear behaviour of single piles using nonlinear Winkler springs and approximating the group stiffness by using static interaction factors. The nonlinear Winkler springs were obtained from load-deflection curves on prototype piles in various soil conditions. The original basic studies were conducted by Matlock (1963), Matlock et al. (1978) and Reese et al. (1974a, 1974b). Their studies led to the p-y procedure for offshore pile design recommended by the American Petroleum Institute (1991). These curves were derived under static conditions or low-frequency cycling. Therefore, they do not incorporate the frequency effect on damping and stiffness. However, in many low frequency applications, the dynamic stiffness is similar to the static stiffness. Nogami et al. (1992) developed a second-order subgrade model based on introducing shear coupling between the distributed springs. It appears to model well the response at all frequencies including the static condition when suitably calibrated.

For seismic analysis of pile foundations under strong shaking, engineering practice relies mainly on the p-y approach. There are well recognized problems with this approach; determining the appropriate p-y curves for the site, the approximate nature of the representation of field conditions, and the difficulty of simulating appropriate dynamic interaction between the piles. Traditionally, static interaction factors have been used, but with their wider availability dynamic factors may be adopted in future. Of course both sets of factors are based on elastic analysis. During nonlinear response, the effective stiffness of the pile group is affected strongly by the load on the group. These inertial effects of the superstructure cause additional strains in the ground and hence modify further the effective moduli and damping. This effect was recognized by Matlock et al. (1978) in the development of the computer program SPASM, based on the p-y curve concept. It will be demonstrated later how important it is to include the forces imposed by the superstructure, when evaluating dynamic pile group stiffness.

A procedure for nonlinear dynamic analysis of pile groups, under development at UBC, will be described here. It is a quasi-3D method which permits dynamic nonlinear analysis of pile groups in layered soils. By relaxing some of the boundary conditions associated with a full 3D analysis, the computing costs can be substantially reduced and the analysis is feasible on a Pentium PC. The procedure is validated here by data from centrifuge tests on a single pile and a 2×2 pile group.

Quasi-3D Lateral Analysis of Piles

Under vertically propagating shear waves (Fig. 1) the soil undergoes primarily shearing deformations in xOy plane except in the area near the pile where extensive compressional deformations develop in the direction of shaking. The compressional deformations also generate shearing deformations in yOz plane. Therefore, the assumptions are made that dynamic response is governed by the shear waves in the xOy and yOz planes, and the compressional waves in the direction of shaking, Y. Deformations in the vertical direction and normal to the direction of shaking are neglected. Comparisons with full 3-D elastic solutions confirm that these deformations are relatively unimportant for horizontal shaking. Applying dynamic equilibrium in Y-direction, the dynamic governing equation under free vibration displacements, v, of the soil continuum is given by

$$\rho_s \frac{\partial^2 v}{\partial t^2} = G * \frac{\partial^2 v}{\partial x^2} + \frac{2}{1-\mu} G * \frac{\partial^2 v}{\partial y^2} + G * \frac{\partial^2 v}{\partial z^2} \tag{1}$$

where $G*$ is the complex shear modulus, ρ_s is the mass density of the soil, and μ is the Poisson's ratio of the soil. A similar equation can be written for vertical motion.

Piles are modelled using ordinary Eulerian beam theory. Bending of the piles occurs only in the yOz plane. Dynamic pile-soil-pile interaction is maintained by enforcing displacement compatibility between the pile and soils.

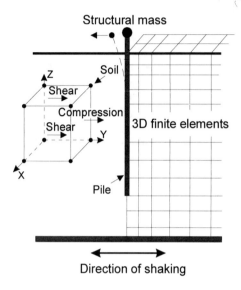

Figure 1. Quasi-3D model of pile-soil response.

A quasi-3D finite element code PILE3D (Wu and Finn, 1994) was developed to incorporate the dynamic pile-soil-pile interaction described previously. An 8-node brick element is used to represent soil, and a 2-node beam element is used to simulate the piles. Using the standard procedure of the finite element method, the global dynamic equilibrium equations are written in matrix form as

$$[M^*]\left\{\frac{\partial^2 v}{\partial t^2}\right\} + [C^*]\left\{\frac{\partial v}{\partial t}\right\} + [K^*]\{v\} = \{P(t)\} \qquad (2)$$

in which P(t) is the dynamic external load vector, and $\{\partial^2 v/\partial t^2\}$, $\{\partial v/\partial t\}$ and $\{v\}$ are the relative nodal acceleration, velocity and displacement, respectively. [M*], [C*] and [K*] are the mass, damping and stiffness matrices of the soil-pile system in the direction of vibration for either horizontal or vertical vibration. [K*] is a complex stiffness matrix which includes the hysteretic damping of the soil.

Direct step-by-step integration using the Wilson-θ method is employed in PILE3D to solve the equations of motion in eqn. (2). In nonlinear analysis, the hysteretic behaviour of soil is modelled by using a variation of the equivalent linear method used in the SHAKE program (Schnabel et al., 1972). Compatibility between shear strains and moduli and damping, is enforced at selected times during shaking, rather than at the end of shaking as in SHAKE. Additional features such as tension cut-off and shearing failure are incorporated in the program to simulate the possible gapping between soil and pile near the soil surface and yielding in the near field.

The loss of energy due to radiation damping is modelled approximately, using the method proposed by Gazetas et al. (1993). A velocity proportional damping force F_d per unit length along the pile is given by

$$F_d = c_x \frac{dv}{dt} \tag{3}$$

where the radiation dashpot coefficient, c_x, is given by

$$c_x = 6\rho_s V_s d \left(\frac{\omega \cdot d}{V_s} \right)^{-0.25} \tag{4}$$

in which V_s is the shear wave velocity of the soil, d is diameter of the pile, and ω is the excitation frequency of the external load. Radiation damping associated with vertical motion of the piles is handled in a similar way using a radiation dashpot coefficient, c_z, with

$$c_z = \rho_s V_s d \left(\frac{\omega \cdot d}{V_s} \right)^{-0.25} \tag{5}$$

Elastic Pile Head Impedances

The impedances K_{ij} are defined as the complex amplitudes of harmonic forces (or moments) that have to be applied at the pile head in order to generate a harmonic motion with a unit amplitude in the specified direction (Novak, 1991). The translational, the cross-coupling, and the rotational impedances of the pile head are represented by K_{vv}, $K_{v\theta}$, $K_{\theta\theta}$, respectively. There is also a vertical stiffness K_{zz}.

Since the pile head impedances are complex valued, they are usually expressed by their real and imaginary parts as,

$$K_{ij} = k_{ij} + i\, C_{ij} \quad \text{...or} \quad K_{ij} = k_{ij} + i\, \omega\, c_{ij} \tag{6}$$

in which k_{ij} and C_{ij} are the real and imaginary parts of the complex impedances, respectively; $c_{ij} = C_{ij}/\omega$ = coefficient of equivalent viscous damping; and ω is the circular frequency of the applied load. k_{ij} and C_{ij} are usually referred to as the stiffness and damping at the pile head. All the parameters in eqn. (6) are dependent on the frequency ω, and, in the case of strong shaking, on the current distribution of effective moduli and damping ratios.

Rocking Impedances of a Pile Group

The rocking impedances of a pile group is a measure of the complex resistance to rotation of the pile cap due only to the resistance of each pile in the group to vertical displacements. The rocking impedance K_{RR} of a pile group is defined as the summation of the moments of the axial pile forces around the centre of rotation of the pile cap for a harmonic rotation with unit amplitude at the pile cap. This definition is quantitatively expressed as

$$K_{RR} = \Sigma r_i \times F_i \tag{7}$$

where r_i are distances between the centre of rotation and the pile head centres, and F_i are the amplitudes of the axial forces at the pile heads.

In the analysis, the pile cap is assumed to be rigid. For a unit rotation of the pile cap, the vertical displacements w_i^P at all pile heads are determined according to their distances from the centre of rotation r_i.

The lateral impedance, K_{vv}, and the rocking impedance, K_{RR}, of a 4-pile group with a pile spacing, $s/d = 5$, where s is the centre to centre spacing, and d is the pile diameter, will now be evaluated using the proposed model, assuming elastic response. The results will be compared with the impedances from complete 3-D analyses to check the computational adequacy of the quasi-3D model. The piles have an L/d ratio >15; the ratio of pile modulus to soil modulus, $E_p/E_S = 1000$; the Poisson's ratio $\mu = 0.4$, and the critical damping ratio, $\lambda = 5\%$. The pile cap is rigid and rigidly connected to the pile heads.

In order to show the pile group effect, dynamic impedances of the pile group are normalized to a stiffness of the pile group expressed as the static stiffness of a single pile times the number of piles in the group. This normalized lateral impedance of the pile group, α_{vv}, is the dynamic interaction factor for lateral loading for the given pile head conditions and is defined as

$$\alpha_{vv} = \frac{K_{vv}}{N \cdot k_{vv}^0} \tag{8}$$

where k_{vv}^0 is the static lateral stiffness of an identical single pile placed in the same soil medium, and N is the number of piles in the pile group. The computed lateral interaction factors for stiffness, α_{vvs}, and damping, α_{vvd}, are compared with those by Kaynia and Kausel (1982) in Fig. 2(a) and Fig. 2(b), respectively. There is good agreement between both sets of factors.

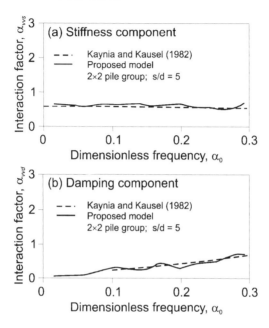

Figure 2. Comparison of dynamic interaction factors for (a) α_{vvs}, and (b) α_{vvd}, with solution by Kaynia and Kausel (1982).

Because the piles are rigidly connected to the pile cap at the pile heads, the total rotational impedance of the pile cap $K_{\theta\theta}^{cap}$ consists of both the rocking impedance K_{RR} of the pile group and the summation of the rotational impedances, $K_{\theta\theta}$, at the head of each pile

$$K_{\theta\theta}^{cap} = K_{RR} + \Sigma\, K_{\theta\theta} \qquad (9)$$

The total rotational impedance of the pile cap $K_{\theta\theta}^{cap}$ is normalized as $K_{\theta\theta}^{cap} / (N \cdot \Sigma r_i^2 k_{zz}^0)$, in which k_{zz}^0 is the static vertical stiffness of an identical single pile placed in the same soil medium. The computed rotational interaction factors for stiffness and damping are compared with those by Kaynia and Kausel (1982) in Fig. 3(a) and Fig. 3(b), respectively. The results obtained using the quasi-3D model are quite satisfactory.

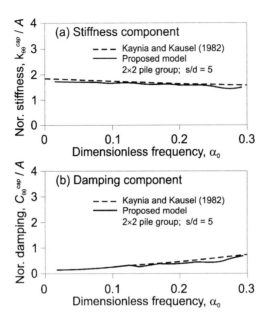

Figure 3. Normalized computed rotational stiffness and damping factors for 2×2 pile group, compared with those of Kaynia and Kausel (1982).

Nonlinear Seismic Response Analysis of a Single Pile

PILE3D was used to analyze the seismic response of a single pile in a centrifuge test conducted at the California Institute of Technology (Caltech) described by Finn and Gohl (1987, 1992) and Gohl (1991). Figure 4 shows the soil-pile-structure system used in the test. The system was subjected to a nominal centrifuge acceleration of 60 g. A horizontal acceleration record with a peak acceleration of 0.158 g is input at the base of the system.

The sand deposit is modelled by 11 layers. Layer thickness is reduced as the soil surface is approached to allow more detailed modelling of the stress and strain field where lateral soil-pile interaction is strongest. The pile is modelled using 15 beam elements including 5 elements above the soil surface. The superstructure mass is treated as a rigid body.

The finite element analysis was carried out in the time domain. Nonlinear analysis was performed to account for the changes in shear moduli and damping ratios due to dynamic shear strains.

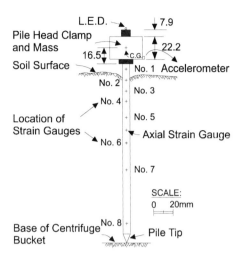

Figure 4. The layout of the centrifuge test for a single pile.

The computed time-history of moments in the pile at a depth of 3 m (near point of maximum moment) is plotted against the recorded time- history in Fig. 5. There is satisfactory agreement between the computed and measured moments in the range of larger moments. The computed and measured moment distributions along the pile at the instant of peak pile head deflection are shown in Fig. 6. The computed moments agree quite well with the measured moments. The moments increase to a maximum value at a depth of 3.5 diameters, and then decrease to zero at a depth around 12.5 diameters. The moments along the pile have same signs at any instant of time, suggesting that the inertial interaction caused by the pile head mass dominates response, and the pile is vibrating in its first mode. The peak moment predicted by the quasi-3D finite element analysis is 344 kNm compared with a measured peak value of 325 kNm.

Seismic Response Analysis of a Pile Group

The seismic response of a 4-pile group in a centrifuge test (Gohl, 1991) was analyzed using the program PILE3D. The piles are set in a 2×2 arrangement at a centre to centre spacing of 2 diameters. The finite element mesh used to analyze the group is shown in Fig. 7. A refined mesh is used around the piles near the surface. This region contributes most to the lateral resistance of the piles, and the shear strains are greatest here. The properties of the piles are identical to those of the single pile described earlier. The group piles were rigidly clamped to a stiff pile cap and four cylindrical masses were bolted to the cap to simulate the inertia of a superstructure.

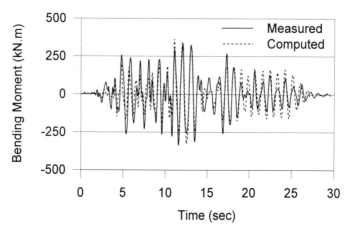

Figure 5. The computed versus measured moment response at depth D = 3 m.

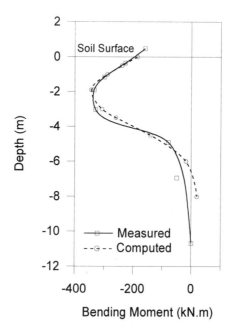

Figure 6. Computed versus measured moment distribution
in the pile at peak pile deflection.

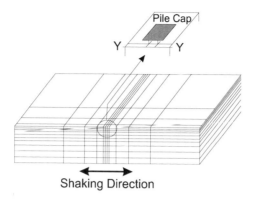

Figure 7. Finite element model of pile group.

At selected times during the horizontal mode analysis, the rocking stiffness and damping is computed using PILE3D in the vertical mode. This impedance calculation is made using the current values of strain dependent moduli and damping. The current rocking impedance is then transferred to the pile cap as rotational stiffness and damping. The accuracy of the representation of rocking impedance depends on the frequency with which it is updated.

The distributions of computed and measured bending moments along the pile at the instant of peak pile cap displacement are shown in Fig. 8. The computed moments agree reasonably well with the measured moments especially in the region of maximum moment. The computed time-history of dynamic moments for the pile group shows the same kind of agreement with the recorded moments as noted for the single pile test.

Pile Impedance in Practice

Commercially available software for the analysis of structures does not incorporate time-dependent stiffnesses. Therefore, for the seismic structural analysis of bridges and buildings on pile foundations, discrete springs and dampers associated with the different degrees of freedom of the pile foundations are assigned to the structural analysis model. The reduction in soil stiffness and the increase in damping associated with strong shaking are sometimes modelled crudely in these analyses by making arbitrary reductions in shear moduli and arbitrary increases in viscous damping. Another approach is to use nonlinear p-y springs and to calculate the stiffness at an arbitrary displacement of the pile head.

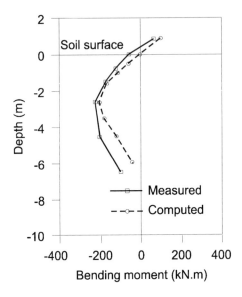

Figure 8. Comparison between measured and computed bending moments
at peak pile cap displacement.

The 3D response analysis program, PILE3D, provides a method for the direct computation of pile impedances taking into account nonlinear behaviour of the soil, pile-soil-pile interaction, hysteretic and radiation damping, and structural inertia effects. This capability allows a more rational selection of effective spring and damping constants for use in commercial software.

The distribution of shear moduli at a depth of 2.1 m in the soil around the pile in Fig. 4 at a time T = 12.58 secs is shown in Fig. 9. The distribution of shear moduli and damping are both time- and space-dependent. This dependence results in a corresponding time-dependence of stiffness and damping of pile foundations. Dynamic impedances as a function of time were computed using the time- and space-dependent nonlinear shear moduli. Harmonic loads with an amplitude of unity were applied at the pile head, and the resulting equations were solved to obtain the complex valued pile impedances. The impedances were evaluated at the ground surface.

The time-dependent dynamic stiffnesses (real parts of the impedances) and associated displacements of the pile are shown in Fig. 10. The dynamic stiffnesses experienced their lowest values between about 10 and 14 seconds, when the largest displacements occurred at the pile head. It can be seen that the lateral stiffness

Figure 9. Distribution of shear moduli around pile at a depth of 2.1 m
at time 12.58 s into earthquake.

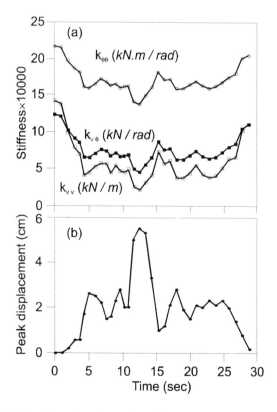

Figure 10. (a) Time-dependent dynamic stiffnesses, and (b) and associated pile-
head displacements for the single pile.

component K_{vv} decreased more than the rotational stiffness $K_{\theta\theta}$ or the coupled lateral- rotational stiffness $K_{v\theta}$. The equivalent damping coefficients (not shown) increased with increasing displacement also because the hysteretic damping of the soil increased with the level of straining.

The time histories of lateral stiffness, K_{vv}, and the coupled lateral-rotational stiffness, $K_{v\theta}$, for the 2×2 pile group subjected to the same input motions as the single pile, are shown in Fig. 11.

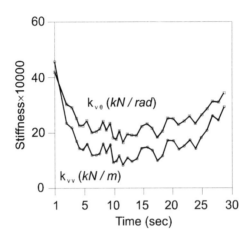

Figure 11. Time-dependent dynamic stiffnesses for the pile group.

The stiffness of a pile group, during seismic shaking, incorporates both kinematic and inertial effects. The inertial effects are due to the mass of the superstructure. Because of the approximate procedures used in practice to evaluate the stiffness of a pile foundation, the inclusion of inertial effects is often ignored. However, the effects of inertial interaction may be very important sometimes.

To demonstrate the effect of inertial interaction, the single pile in Fig. 4 was analyzed, with and without the structural mass at the pile head. The variations in lateral stiffness, with and without inertial interaction, are shown in Fig. 12. In this case, the inertial interaction has a major effect on the stiffness of the pile. It is clear that inertial interaction should be considered when evaluating pile stiffnesses under strong earthquake shaking.

The time histories of lateral and rocking stiffnesses shown in Figs. 10, 11 and 12, show clearly the difficulties in selecting a single spring value to represent the lateral or rotational stiffness of a pile foundation. To make a valid selection,

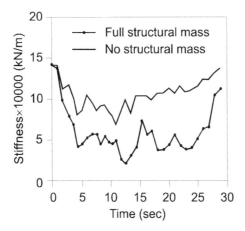

Figure 12. The effect of the inertia of structural mass on time-dependent
foundation stiffness under strong shaking.

one would need to know which segment of the ground motion was most critical in
controlling the seismic response of the structure. A spring based on the minimum
lateral stiffness would represent the mobilized stiffness during the period of very
strong shaking and would be more critical for longer period structures. Knowing
the time variation in stiffness and damping makes it possible to estimate better the
appropriate effective discrete stiffness and damping for use in a commercial
structural analysis program.

Effects of Seismic Porewater Pressures

Porewater pressures are generated by two mechanisms: shearing stresses in
the free field and additional pressure developed by the interaction between soils and
piles. The free field porewater pressures may be estimated using current
technology such as that outlined by Harder and Seed (1990), or dynamic effective
stress analysis using programs such as DESRA-2 (Lee and Finn, 1978), or
Dynaflow (Prevost, 1981). An extension to PILE3D is under development to
incorporate a porewater pressure model and the capability for dynamic effective
stress analyses. In the meantime, the PILE3D analysis is conducted after the free-
field porewater pressures have been evaluated and their effects on soil stiffnesses
have been taken into account in the specification of soil properties.

Foundation Stiffnesses of the Painter Street Bridge Bent Using Pile3D

The Painter Street Overpass located near Rio Dell in Northern California, is a two-span, prestressed concrete box-girder bridge that was constructed in 1973 to carry traffic over the four-lane US Highway 101 (Fig. 13). The bridge is 15.85 m wide and 80.79 m long. The deck is a multi-cell box girder, 1.73 m thick and is supported on monolithic abutments at each end and a two-column bent that divides

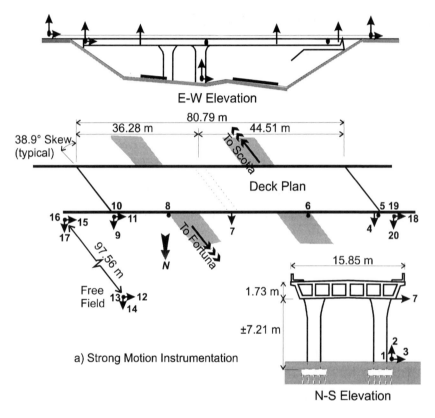

Figure 13. Dimensions and instrumentation of Painter Street Overpass.

the bridge into two spans of unequal length. One of the spans is 44.51 m long and the other is 36.28 m long. The east and west abutments are supported by 14 and 16 piles, respectively. Longitudinal movement of the west abutment is allowed by means of a thermal expansion joint. Each column of the bent is 7.32 m high and supported by 20 concrete friction piles in a 4×5 group.

The bridge was instrumented in 1977 as part of a collaborative effort between the California Strong Motion Instrumentation Program (CSMIP) in the Division of Mines and Geology and CALTRANS to record and study strong motion records from selected bridges in California.   Twenty strong motion accelerometers were installed on and off the bridge as shown in Fig. 13.   The instrumentation has recorded several earthquakes since its installation including the main shock of the Cape Mendocino-Petrolia earthquake of 25 April 1992.   This earthquake had a Richter magnitude M_L = 6.9, and occurred 6.4 km from the bridge. It generated a free-field peak acceleration in the ground near the bridge, a_p = 0.54 g, and a peak acceleration in the bridge structure, a_p = 1.09 g.   Despite the large structural accelerations, no significant structural damage has been observed at the bridge. The extent of damage has been limited to settlement of the backfills and minor spalling of the concrete.

Finn et al. (1995) described studies of the seismic behaviour of the abutments and centre bent of this bridge including system identification analyses and quasi-3D analyses of the pile foundation. Frequency analysis of the bridge deck accelerations under earthquake ground motions indicated that the deck moved as a rigid body in horizontal translation and rotation. Using a modified version of the deck and force system used previously by Goel and Chopra (1995), as shown in Fig. 14, the variation of abutment stiffness during strong shaking was determined as a function of relative displacement between the abutment and the free-field. The variation of stiffness parallel to the face of the east abutment is shown in Fig. 15. It is clear that quite modest displacements bring about a large reduction in the stiffness. The stiffness during strong shaking corresponding to relative displacements greater than about 1 cm is about one-fourth (1/4) of the stiffness under the initial weak excitation by the earthquake. This reduction in stiffness is very similar to that determined by Goel and Chopra (1995).

Changes in dynamic stiffness are reflected in changes in the fundamental period of the bridge. Ventura et al. (1995), from frequency analysis of the records from ambient and strong motion data, showed that under strong shaking the fundamental period of this bridge dropped by 10%. It is clearly desirable to have a reliable method for the direct estimation of stiffness and damping under strong shaking.

Evaluation of Pile Foundation Stiffnesses

The model of the foundation soils was based on data from field tests conducted by the Lawrence Livermore Laboratory in the U.S., and was supplied by Heuze (1994). The shear modulus of the soil varied parabolically with depth with an average value of 100 MPa in the surficial soil layer.

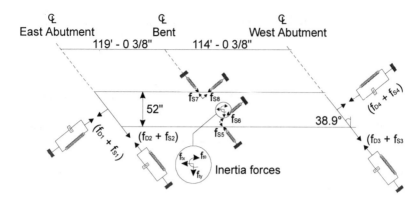

Figure 14. Rigid body model and force system for Painter Street Overpass
(after Gohl and Chopra, 1995).

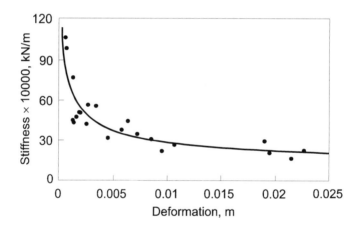

Figure 15. Variation of lateral stiffness parallel to the east abutment
with displacement.

The bridge was subjected to the accelerations of the main shock of the Cape Mendocino-Petrolia earthquake of 1992 with magnitude M = 6.9.

The elastic stiffnesses of the pile foundation are not affected by the inertia of the superstructure because the moduli are not dependent on displacements. The stiffnesses of the real soil under strong shaking are affected by the inertia of the superstructure because the stiffnesses are functions of displacements. Therefore, the proportion of the inertial mass of the deck structure adopted by Makris et al. (1994) in their analysis of the bridge was included in the nonlinear dynamic response of the bridge bent.

The four pile cap stiffnesses of the 20-pile foundation of the bridge bent were evaluated for both elastic and nonlinear response, in order to demonstrate the importance of nonlinear effects. The elastic stiffness of a single pile was also determined to allow an estimate of the effect of pile-to-pile interaction on elastic stiffness under dynamic conditions.

The elastic stiffnesses of a single pile are given in column 2 of Table 1. The elastic stiffnesses of the 20 pile foundation are given in column 3. These stiffnesses are about 50%-30% of the stiffnesses corresponding to 20 times the stiffness of a single pile. This reduction in stiffness is due to pile-soil-pile interaction under elastic dynamic conditions.

The time-history of lateral stiffness under the strong shaking taking nonlinear soil behaviour into account is shown in Fig. 16. It is typical of the variations in all stiffnesses. What discrete single valued spring stiffness should be selected for use in a commercial structural analysis program that adequately represents this time-dependent stiffness? This is obviously a matter of judgement. The authors' choice, based on the time histories of the four stiffnesses are given in column 4 of Table 1. The selected stiffnesses are biased towards the values associated with the time of strongest shaking. A comparison of columns 3 and 4 shows that the stiffnesses based on initial in-situ values of moduli were reduced by factors of 3.5 for lateral stiffness and by 3.2 for rocking stiffness by strong shaking. These reductions are of the same order as those deduced from the system identification analysis of the abutment of the bridge during the Cape Mendocino earthquake by Finn et al. (1995), which are shown in Fig. 15.

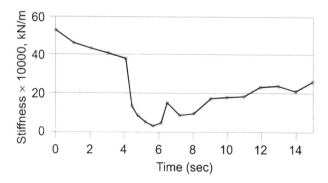

Figure 16. Time variation of lateral stiffness of pile group during strong shaking.

Table 1. Dynamic Stiffnesses of a Single Pile and the 20-Pile Foundation Group

Type of Stiffness	Single Pile (elastic response)	20-Pile Group (elastic response)	20-Pile Group* (strong shaking)
Lateral, k_{uu}	77.7 MN/m	520 MN/m	150 MN/m
Cross Coupling, $k_{u\theta}$	36.7 MN/rad	330 MN/rad	170 MN/rad
Rocking, k_{RR}	--	8400 MN.m/rad	2600 MN.m/rad
Vertical, k_{zz}	376 MN/m	3700 MN/m	2300 MN/m

*Effective stiffness selected for use in commercial structural program based on time histories.

Conclusions

A validated method for elastic and nonlinear dynamic analysis of pile foundations based on a simplified 3D model of the half space called PILE3D has been presented which can calculate the time-histories of stiffness (and damping) of a pile foundation during an earthquake. The analysis can also include the effects of superstructure inertia on both stiffness (and damping). The time variation of these parameters for a given design earthquake allows a more realistic selection of the representative discrete stiffnesses and damping ratios required by structural analysis programs than the rather arbitrary procedures often used in practice.

Acknowledgements

Research on seismic soil-structure interaction is supported by grants to the lead author by the Natural Science and Engineering Council of Canada.

References

American Petroleum Institute. (1991). "Recommended Practice for Planning, Designing and Constructing Fixed Offshore Platforms," 19th Ed., Washington, DC.

Davies, T.G., R. Sen, and P.K. Banerjee. (1985). "Dynamic Behaviour of Pile Groups in Inhomogeneous Soil," J. Geot. Eng., ASCE, Vol.111, No.12, pp.1365-1379.

El-Masrafawi, H., A.M. Kaynia and M. Novak. (1992a). "Interaction Factors and the Superposition Method for Pile Group Dynamic Analysis," Research Report, GEOT-1-1992, University of Western Ontario, London, Ontario.

El-Masrafawi, H., A.M. Kaynia and M. Novak. (1992b). "The Superposition Approach to Pile Group Dynamics," Geot. Special Publication No.34, ASCE, New York, N.Y., pp.114-135.

Finn, W.D. Liam and W.B. Gohl. (1987). "Centrifuge Model Studies of Piles Under Simulated Earthquake Lateral Loading," in "Dynamic Response of Pile Foundations - Experiment, Analysis and Observation", Toyoaki Nogami (Ed.), ASCE, Geotechnical Special Publication No. 11, pp. 21-38.

Finn, W.D. Liam and W.B. Gohl. (1992). "Response of Model Pile Groups to Strong Shaking," in "Piles Under Dynamic Loads", Shamsher Prakash (Ed.), ASCE, Geotechnical Special Publication No. 34, pp. 27-55.

Finn, W.D. Liam and G. Wu. (1994). "Recent Developments in Dynamic Analysis of Piles," Proc., 9th Japan Conf. on Earthq. Eng., Tokyo, Japan, Vol.3, pp.325-330.

Finn, W.D. Liam, G. Wu and T. Thavaraj. (1995). "Seismic Response for Pile Foundations for Bridges," Proc., 7th Can. Conf. on Earthq. Eng., pp.779-786, May.

Gazetas, G. and M. Makris. (1991a). "Dynamic Pile-Soil-Pile Interaction. Part I: Analysis of Axial Vibration," Earthq. Eng. Struct. Dyn., Vol.20, No.2, pp.115-132.

Gazetas, G., M. Makris and E. Kausel. (1991b). "Dynamic Interaction Factors for Floating Pile Groups," J. Geot. Eng., ASCE, Vol.117, No.10, pp.1531-1548.

Gazetas, G., K. Fan and A. Kaynia. (1993). "Dynamic Response of Pile Groups with Different Configurations", Soil Dynamic & Earthq. Eng., Vol.12, pp.239-257.

Gohl, W.B. (1991). "Response of Pile Foundations to Simulated Earthquake Loading: Experimental and Analytical Results," Ph.D. Thesis, Dept. of Civil Eng., University of British Columbia, Vancouver, B.C., Canada.

Heuze, F.E. (1994). Private Communication.

Kaynia, A.M. (1982). "Dynamic Stiffness and Seismic Response of Pile Groups," Research Report, R82-03, Dept. of Civil Eng., Cambridge, Mass.

Lee, M.K.W. and Finn, W.D.L. (1978). "DESRA-2, Effective Stress Response Analysis of Soil Deposits with Energy Transmitting Boundary Including Assessment of Liquefaction Potential," Soil Mechanics Series No. 38, Department of Civil Engineering, University of British Columbia, Vancouver, B.C.

Makris, N., D. Badoni, E. Delis and G. Gazetas. (1994). "Prediction of Observed Bridge Response with Soil-Pile-Structure Interaction", J. Structural Eng., ASCE, Vol. 120, No.10, October.

Matlock, H. (1963). "Applications of Numerical Methods to Some Structural Problems in Offshore Operations," J. of Petroleum Tech., September.

Matlock, H., S.H.C. Foo, and L.M. Bryant (1978). "Simulation of Lateral Behaviour Under Earthquake Motion," Proc., Geot. Div. Specialty Conf. on Earthq. Eng. & Soil Dyn., Amer. Soc. of Civil Engrs., Pasadena, CA, June, pp. 601-619.

Nogami, T., J-X. Zhu and T. Ito (1992). "First and Second Order Dynamic Subgrade Models for Soil-Pile Interaction Analysis", in "Piles Under Dynamic

Loads", Shamsher Prakash (Ed.), ASCE, Geotechnical Special Publication No. 34, pp. 187-206.

Novak, M. (1991). "Piles Under Dynamic Loads," State of the Art Paper, 2nd Int. Conf. Recent Advances in Geot. Earthq. Eng. and Soil Dynamics, University of Missouri-Rolla, Rolla, Missouri, Vol.III, pp. 250-273.

Prevost, J.H. (1981). "DYNAFLOW: A Nonlinear Transient Finite Element Analysis Program," Department of Civil Engineering, Princeton University, Princeton, New Jersey.

Reese, L.C., W.R. Cox, and F.D. Koop (1974a). "Analysis of Laterally Loaded Piles in Sand," 6th Annual Offshore Tech. Conf., Houston, TX, May, Paper #2080.

Reese, L.C., W.R. Cox and F.D. Koop (1974b). "Field Testing and Analysis of Laterally Loaded Piles in Stiff Clay," 7th Annual Offshore Tech. Conf., Houston, TX, May, Paper No.2312.

Schnabel, P.B., J. Lysmer and H.B. Seed. (1972). "SHAKE: A Computer Program for Earthquake Response Analysis of Horizontally Layered Sites," Report EERC 71-12, University of California at Berkeley.

Ventura, C.E., W.D. Liam Finn and A.J. Felber. (1995). "Dynamic Testing of Painter Street Overpass", Proc., 7th Canadian Conf. on Earthquake Eng., pp. 787-794, June.

Wu, G. (1994). "Dynamic Soil-Structure Interaction: Pile Foundations and Retaining Structures," Ph.D. Thesis, Dept. of Civil Eng., University of British Columbia, Vancouver, B.C., Octobert, 198 pgs.

Wu, G. and W.D. Liam Finn. (1994). "PILE3D - Prototype Program for Nonlinear Dynamic Analysis of Pile Groups" (still under development). Dept. of Civil Eng., University of British Columbia, Vancouver, B.C.

Wu, G. and W.D. Liam Finn (1995). "A New Method for Dynamic Analysis of Pile Groups," Proc., 7th Int. Conf. on Soil Dyn. & Earthq. Eng., Chania, Crete, Greece, May 24-26, Vol.7, pp.467-474.

ANALYSIS OF R/C CHIMNEYS WITH SOIL-STRUCTURE INTERACTION

Jon K. Galsworthy[1] and M. Hesham El Naggar[2], Member, ASCE

ABSTRACT

Reinforced concrete chimneys are often analyzed as fixed base cantilever beams ignoring the effect of soil-structure interaction on the response. However, the foundation flexibility has a significant effect on their response to earthquake, wind and other dynamic loads. It influences the dynamic properties of concrete chimneys in two ways; alters the natural frequencies, damping ratios and mode shapes; results in coupling of the response of the liner and the shell. In this study, the stack was modeled as a continuous cantilever beam with distributed stiffness and mass supported by a flexible foundation. Natural frequencies and mode shapes are determined using the Rayleigh-Ritz method. The effect of the liner is considered in the analysis and thus accounting for the possible coupling of vibration modes. The effect of the foundation flexibility on the dynamic properties of the tall chimneys was evaluated for three different types of foundation. It was found that ignoring the foundation flexibility leads to overestimating the natural frequencies by 10 to 20% and hence some errors in mode shapes. It was found that foundation flexibility increases the shell deflection, represented by mode shapes, by almost 100% in liner mode and 20% in the shell mode. Similar observations were made for liner deflection but to a lesser extent. The modal damping ratios increase due to soil-structure interaction, but the increase is small because the large mass of the structure. Neglecting the foundation flexibility may lead to severe errors in the prediction of the response to dynamic loads.

INTRODUCTION

There are several different structural forms for reinforced concrete chimneys. One common form is a free standing brick liner enclosed by a concentric reinforced concrete shell, with both supported by a common foundation. This form has been

---

[1] Research Assistant, Faculty of Engineering Science, The University of Western Ontario, London, Ontario. N6A 5B9

[2] Assistant Professor, Faculty of Engineering Science, The University of Western Ontario, London, Ontario. N6A 5B9

observed to be sensitive to certain aerodynamic conditions (Vickery, 1995). The conventional method of analysis, that assumes a fixed base cantilever, fails to predict the relative liner/shell motions caused by coupling through the foundation. This coupling can lead to impact at the tip causing failure of the liner leading to possible corrosive attack of the concrete shell. The correct prediction of this motion depends on the accuracy of description of the foundation and structural properties.

The soil-structure interaction has two main effects on the mode shapes. First, curvatures in both the shell and liner are reduced due to rotation of the foundation. Second, the liner is excited by this same rotation. The former leads to a reduction in structural damping. Depending on the underlying soil, this may or may not be compensated by increased foundation damping (Novak and El Hifnawy, 1983). The present paper examines how the response is affected by changes in foundation type and soil properties.

FOUNDATION MODEL

Chimneys could be supported by different types of foundations. The types most frequently encountered are shallow foundations and deep foundations. Deep foundations are a group of piles supporting a rigid pile cap (Fig.1a). Piles could be extending in a deep homogeneous soil deposit (floating piles) or penetrating a soil deposit and resting on the bedrock (end-bearing piles). Shallow foundations are rigid blocks partially or fully embedded in the soil (Fig. 1b).

The foundation dynamic stiffness is a complex function whose real part represents the true stiffness, defined by the stiffness constant, k, and its imaginary part represents damping, defined by the constant of equivalent viscous damping, c, i.e.

$$K = k + i \, c \qquad\qquad (1)$$

The stiffness of a shallow foundation stems from two components; base soil reaction and side soil reaction. The base soil reaction is calculated using the results of Wong and Luco (1985). This solution is exact for square footings. For circular footing, the stiffness is evaluated approximately using equivalent dimensions obtained by equating the geometric properties of the base area of the actual footing with those of a square base. The side soil reaction (the embedment effect) is calculated using the approach due to Beredugo and Novak (1972).

The stiffness of the deep foundation is derived using the superposition approach described in El Naggar and Novak (1995) in which the stiffness of individual piles is considered first and then the group effect is introduced through the interaction factors approach. The stiffness of the individual piles is evaluated by the plane strain approach given by Novak and Aboul-Ella (1978a, b) while the interaction factors are derived based on the charts of Kaynia and Kausel (1982).

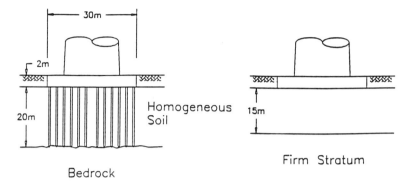

Fig. 1a: End-Bearing Pile Foundation   Fig 1b: Shallow Foundation

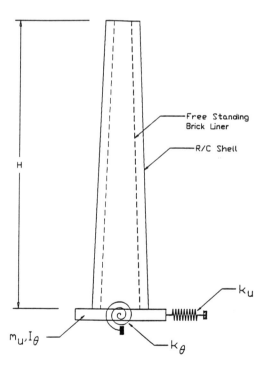

Fig. 1c: Model for Soil-Structure Interaction

The governing equation of undamped free vibration for the foundation in the horizontal translation and rocking vibration modes can then be given by

$$\begin{bmatrix} I_f & 0 \\ 0 & m_f \end{bmatrix} \begin{Bmatrix} \ddot{\theta} \\ \ddot{u} \end{Bmatrix} + \begin{bmatrix} k_{\theta\theta} & k_{\theta u} \\ k_{u\theta} & k_{uu} \end{bmatrix} \begin{Bmatrix} \theta \\ u_f \end{Bmatrix} = \{0\} \tag{2}$$

where u_f and θ are the translation and rotation of the foundation, respectively, m_f and I_f are its mass and mass moment of inertia, respectively.

Soil-Structure Interaction Model

The general structural arrangement is shown in Fig. 1c. The super-structure consists of two, thin-walled circular beams subject to lateral loads. It is assumed that shear deformations and negative geometric stiffness can be ignored. The former is valid for aspect ratios (height/diameter) greater than 10 and structural stability is paramount. The cross section is assumed to be stiff in the radial and circumferential directions, thus eliminating ovalling. The foundation is considered rigid with degrees of freedom in the rocking and translating directions. Due to the symmetry of the structure, torsional motions are not considered.

The Rayleigh-Ritz method is adopted for the free-vibration analysis. The shell and liner are treated as beam elements with distributed mass per unit length and flexural rigidity with taper in diameter and/or wall thickness. The shape functions used are the true mode shapes for a uniform cantilever beam for elastic deformations and rigid body rotation and translation for the foundation. Consistent mass and stiffness matrices can then be calculated using the principles of virtual work (Clough and Penzien, 1993). The undamped free vibration equations of motion in matrix form are

$$\begin{bmatrix} [M_s] & [0] & \{M_{s_\theta}\} & \{M_{s_u}\} \\ [0] & [M_l] & \{M_{l_\theta}\} & \{M_{l_u}\} \\ \{M_{s_\theta}\}^T & \{M_{l_\theta}\}^T & I_f + m_{\theta\theta} & m_{\theta u} \\ \{M_{s_u}\}^T & \{M_{l_u}\}^T & m_{u\theta} & m_f + m_{uu} \end{bmatrix} \begin{Bmatrix} \{\ddot{a}\}_s \\ \{\ddot{a}\}_l \\ \ddot{\theta} \\ \ddot{u}_f \end{Bmatrix}$$

$$+ \begin{bmatrix} [K_s] & [0] & \{0\} & \{0\} \\ [0] & [K_l] & \{0\} & \{0\} \\ \{0\}^T & \{0\}^T & k_{\theta\theta} & k_{\theta u} \\ \{0\}^T & \{0\}^T & k_{u\theta} & k_{uu} \end{bmatrix} \begin{Bmatrix} \{a\}_s \\ \{a\}_l \\ \theta \\ u_f \end{Bmatrix} = \{0\} \tag{3}$$

In Equation 3, the submatrices $[M_s]$ and $[M_l]$ are the mass matrices for the shell and liner, respectively, and $[K_s]$ and $[K_l]$ are the stiffness matrices for the shell and liner, respectively. These matrices have dimensions $n \times n$ where n is the number of elastic shape functions employed. The null vector, $\{0\}$, has n zero elements and indicates no contribution to the elastic stiffness from the foundation. The other terms appearing in Equation 3 are

$$M_{s_{\theta_i}} = \int_0^H m_s(z) \cdot \frac{z}{H} \cdot \psi_i(z)dz \qquad M_{l_{\theta_i}} = \int_0^H m_l(z) \cdot \frac{z}{H} \cdot \psi_i(z)dz \qquad (4)$$

$$M_{s_{u_i}} = \int_0^H m_s(z) \cdot 1 \cdot \psi_i(z)dz \qquad M_{l_{u_i}} = \int_0^H m_l(z) \cdot 1 \cdot \psi_i(z)dz \qquad (5)$$

$$m_{\theta\theta} = \int_0^H \left(m_s(z) + m_l(z) \right) \cdot \left(\frac{z}{H} \right)^2 dz \qquad (6)$$

$$m_{\theta u} = m_{u\theta} = \int_0^H \left(m_s(z) + m_l(z) \right) \cdot 1 \cdot \frac{z}{H} dz \qquad (7)$$

$$m_{uu} = \int_0^H \left(m_s(z) + m_l(z) \right) \cdot 1 \cdot 1 dz \qquad (8)$$

$$k_{\theta\theta} = \frac{(k_\theta)_f}{H^2}, \quad k_{u\theta} = k_{\theta u} = \frac{(k_{u\theta})_f}{H}, \quad k_{uu} = (k_u)_f. \qquad (9)$$

Equation 3 defines the eigenvalue problem where the eigenvectors form the coefficients for the shape functions used in assembling the mass and stiffness matrices. The undamped mode shapes for the liner and shell for the jth mode is,

$$\phi_{s_j} = \left(\sum_{i=1}^n a_{i,j} \cdot \psi_i \right)_s + \left(\theta_j \cdot \frac{z}{H} + u_j \right)_f$$

$$\phi_{l_j} = \left(\sum_{i=n+1}^{2n} a_{i,j} \cdot \psi_{i-n} \right)_l + \left(\theta_j \cdot \frac{z}{H} + u_j \right)_f \qquad (10)$$

where $a_{i,j}$ = coefficient in eigenvector corresponding to the ith elastic shape function, ψ_i, in the jth mode, θ_j, u_j = foundation coefficients for jth mode. The solution is calculated using the Stodola - Vianello method with the mode shapes normalized to a unit total tip deflection for the shell.

Due to low observed total damping in full scale chimneys, typically < 1% of critical (Basu, 1982), the undamped mode shapes are accepted as the true shapes. Also, the natural frequencies are well separated thus a proportional damping matrix is a reasonable approximation.

FREE VIBRATION RESULTS

A real structure for which data are available is analyzed to demonstrate the effect of foundation flexibility. It has a height of 250 m with an aspect ratio (height/diameter) of the shell equal to 15.7. Different foundation types are included in the analysis. The soil shear wave velocity is varied from 100 m/s to 250 m/s to represent typical values of real soils.

For comparison, the fixed foundation results are calculated. Using n=3 in Equation 3, the first natural frequencies of the liner and shell are .22 Hz and .28 Hz respectively. The liner mode shape differs only slightly from that of a uniform cantilever with approximately a 6% contribution from the second elastic shape function and <1% from the third. Similar results are observed for the shell. No change was observed by adding more functions.

End-Bearing Piles

The stiffness of a foundation with 250 end-bearing piles is calculated for the range in soil shear wave velocity. In all cases, the unit weight of the soil γ =18 kN/m^3 and Poisson's ratio v = 0.3. The shear wave velocity for the bedrock is taken as 500 m/s with γ and v being the same. The foundation consists of 6 rings of HP360x132 (ASTM HP14x89) piles with uniform centre to centre spacing of 4 diameters (~1.4m) with a rigid cap approximately 30 m in diameter.

The first two modes of the chimney are evaluated. Natural frequencies are reduced from the liner and shell modes based on a fixed foundation. For the first mode, henceforth called the liner mode, the liner and shell move in phase with tip amplitudes for the liner exceeding the shell. In the second, or shell mode, the liner and shell are antiphase with shell tip deflections exceeding (except for very flexible foundations) those of the liner. The translation of the foundation contributes very little (<1%) to the total deflections, for the range of soil parameters considered, and is ignored. However this may not be true for chimneys with lower aspect ratios and chimneys where the elastic frequencies are closer to that of the foundation.

The results for the chimney supported by the end-bearing piles foundation are shown in Figs. 2-5. The rocking stiffness of the foundation increases by a factor of 2.1 over the range in V_s. The frequency for the liner mode is affected the most, as shown in Fig. 2, with a reduction of 20% at V_s=100 m/s and 8% at V_s=250 m/s. The

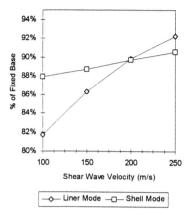

Fig. 2: Comparison of Natural Frequencies for End-Bearing Pile Foundation

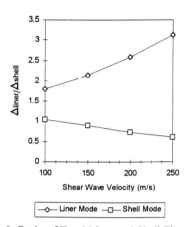

Fig. 3: Ratio of Total Liner and Shell Tip Deflections

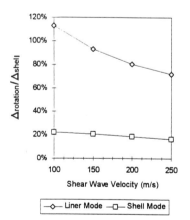

Fig. 4: Comparison of Rocking and Elastic Tip Deflections of the Shell

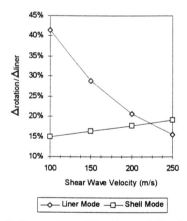

Fig. 5: Comparison of Rocking and Elastic Tip Deflections of the Liner

corresponding reductions in frequency for the shell mode are 12% and 9.5%. The total liner deflection as a ratio of the total shell deflection is shown in Fig. 3. As the rocking stiffness increases, the mode shapes tend towards the fixed foundation case with the liner dominating the liner mode and the same for the shell in the shell mode. Figures 4 and 5 show the ratio of the deflection due to the rotation of foundation to the fixed-base deflections in the liner and shell modes. In the liner mode, shell deformations are dominated by rigid body rotation. This effect decreases with increasing the foundation stiffness but remains as the dominant deformation. The same trends are present for the liner but to a lesser extent because of the higher relative stiffness. In the shell mode, the foundation has a lesser effect for both the liner and shell. The trend of increasing deflection for the liner is because of the mode is anti-symmetry and indicates reduced rocking.

Floating Piles

The floating piles foundation is identical to the end-bearing case, but the soil is assumed to be a deep homogeneous deposit. Soil properties are the same as for the end-bearing case. Foundation rocking stiffness increases by a factor of 3.4 over the range in V_s, attributable to lack of the bedrock contribution. The relative values increase from 60% to 90% of the end-bearing foundation. The results for the chimney supported by floating piles are shown in Figs. 6-9. The changes in natural frequency, depicted in Fig. 6, are similar to the end-bearing case. For the liner mode, the frequency varies from 70% to 90% of the fixed values. Figure 7 shows that the relative liner/shell deflections vary from the case where the two modes are virtually the same (180 degrees out of phase) to similar values seen in the end-bearing case. The ratio of foundation to elastic deflections is sensitive to slight changes in rocking stiffness as seen in Figs. 8 and 9. The 40% decrease in rocking stiffness causes a 70% increase in the ratio of foundation to elastic shell deflections in the liner mode. The ratio approaches the end-bearing case with a difference of 3% caused by a 10% change in rocking stiffness. Similar trends are observed for the liner. This emphasizes the importance of accurate input data for a dynamic analysis. Small changes in elastic moduli of the concrete and brick, and the shear wave velocity of the soil can alter the relative foundation contribution to the generalized stiffness which can lead to large errors in the mode shape estimation thus relative liner/shell deformations.

Shallow Foundation

The configuration of the shallow foundation is shown in Figure 1b. Dimensions of the foundation are the same as the rigid pile cap. The underlying soil is homogeneous down to a firm stratum (V_s=400 m/s). Foundation stiffness is evaluated for a ratio of stratum depth to foundation radius equal to 1.0 and ν=.33. The soil density is the same as the previous cases.

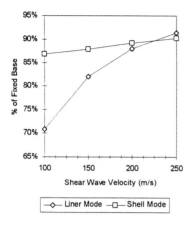

Fig. 6: Comparison of Natural Frequencies for Floating Pile Foundation

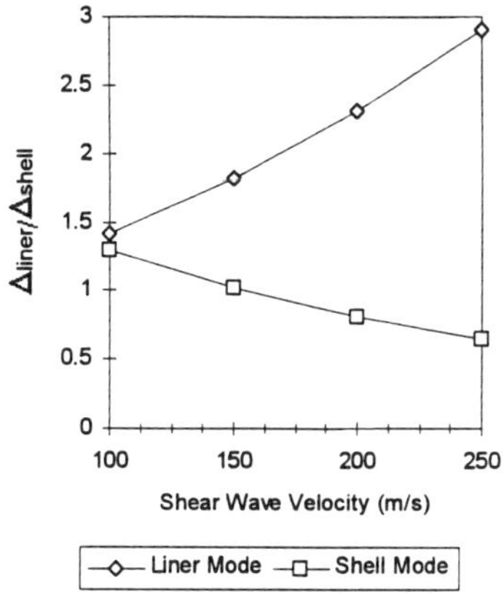

Fig. 7: Ratio of Total Liner and Shell Tip Deflections

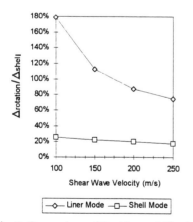

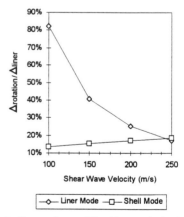

Fig. 8: Comparison of Rocking and Elastic Tip Deflections of the Shell

Fig. 9: Comparison of Rocking and Elastic Tip Deflections of the Liner

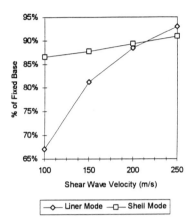

Fig. 10: Comparison of Natural Frequencies for Shallow Pile Foundation

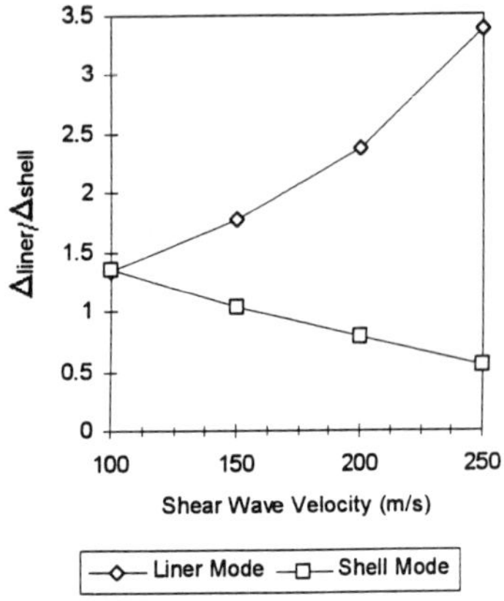

Fig. 11: Ratio of Total Liner and Shell Tip Deflections

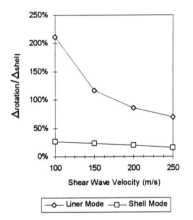

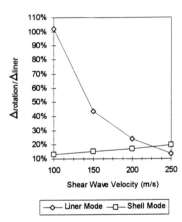

Fig. 12: Comparison of Rocking and Elastic Tip Deflections of the Shell

Fig. 13: Comparison of Rocking and Elastic Tip Deflections of the Liner

Results are shown in Figs. 10-13. The rotational stiffness increases by a factor of 4.9 over the range in V_s. It increases from 50% to 110% relative to the end-bearing pile case, and 80% to 120% of the floating pile case. The change in natural frequencies, shown in Fig. 10, is similar to the previous cases in each mode. The relative liner/shell deflection, depicted in Fig. 11, is similar to both the floating and end-bearing pile cases with the two modes approaching identical shapes at low values of V_s. At higher values of V_s, the change in liner deflection is sensitive to small changes in foundation stiffness, with roughly a 10% increase in relative tip deflections with a 20% increase in rotational stiffness over the end-bearing case. The same trends observed in the relative rotation/elastic deflections of the deep foundation case can be noticed in Figs. 12 and 13.

FOUNDATION DAMPING

Foundation damping is calculated based on the complex impedance functions used in evaluating the frequencies and mode shapes. The complex part is interpreted as an equivalent coefficient of viscous damping. The undamped mode shapes are used to estimate the modal damping through an energy approach They represent a good approximation and require much less computational effort than the mathematically accurate complex eigenvalue approach (Novak and El Hifnawy, 1983).

$$\beta_f = \frac{c}{c_{crit}} = \frac{c_{\theta\theta} \cdot \left(\frac{\theta_j}{H}\right)^2 + 2 \cdot c_{u\theta} \cdot u_{f_j} \cdot \left(\frac{\theta_j}{H}\right) + c_{uu} \cdot u_{f_j}}{2 \cdot M^*_j \cdot \omega_j} \quad (11)$$

In the calculation of the impedance functions, material damping $\tan(\delta)=0.1$ where

$$G^* = G(1 + i\tan\delta) \quad (12)$$

is the complex shear modulus calculated from the hysteretic loop by the following

$$\tan \delta = \frac{1}{2\pi} \cdot \frac{\Delta W}{W} \quad (13)$$

in which, ΔW = the area enclosed by the loop and W = the strain energy.

The results for the three foundation types considered in this study are shown in Table 1. Both $c_{u\theta}$ and c_{uu} are ignored in calculating β_f, with observed errors <1%.

Table 1 Foundation Modal Damping, β_f

V_s	End-Bearing		Floating		Shallow	
(m/s)	Liner	Shell	Liner	Shell	Liner	Shell
100	0.35%	0.07%	0.27%	0.03%	0.85%	0.06%
150	0.30%	0.09%	0.25%	0.05%	0.68%	0.13%
200	0.25%	0.10%	0.22%	0.07%	0.49%	0.18%
250	0.20%	0.11%	0.19%	0.09%	0.32%	0.20%

The larger values of damping in the liner mode are due to the lower mass per unit length of the liner resulting in a lower generalized mass. Increasing V_s decreases modal damping in the liner mode with the opposite being true for the shell mode. In the liner mode, the decrease in damping is due to the large decrease in the foundation rotation with increasing V_s. A less significant decrease in the foundation contribution is present in the shell mode. Thus, the slight increase in damping with V_s is due to increased radiation damping.

The low values of damping observed in Table 1 are not unexpected. For a massive structure like a reinforced concrete chimney, the large generalized mass of the structure limits the foundation damping. Thus it is reasonable to make an estimate of overall damping in calculating response. This may not be the case for lighter structures such as steel chimneys where the foundation makes a larger contribution to damping 0and might be exploited to mitigate aerodynamic instabilities like galloping.

CONCLUSIONS

The modal properties of reinforced concrete chimneys supported by three different types of foundation were evaluated considering soil-structure interaction. The soil-structure interaction alters the natural frequencies, mode shapes and modal damping ratios. Natural frequencies decrease and mode shapes increase for both the shell and liner. Modal damping ratios increase due to soil-structure interaction , but the increase is generally small because the large mass of the structure.

REFERENCES

Basu, R., 1982. Across-Wind Response of Slender Structures of Circular Cross-Section to Atmospheric Turbulence. Ph.D. Thesis, The University of Western Ontario, Faculty of Engineering Science, London, Canada.

Beredugo, Y. O. and Novak, M. 1972. Coupled horizontal and rocking vibration of embedded footings. Canadian Geotechnical Journal, Vol. 9, No. 4, pp. 477-497.

Clough, R.W. and Penzien, J. 1993. *Dynamics of Structures*, McGraw-Hill, New York.

El Naggar, M.H. and Novak, M., 1995. Nonlinear lateral interaction in pile dynamics. International Journal of Soil Dynamics and Earthquake Engineering, Vol. 14, No. 2, pp. 141-157.

Kaynia, A. M. and Kausel, E., 1982. Dynamic Behaviour of pile groups. Conference on Numerical Methods in Offshore Piling, University of Texas, Austin, TX, pp. 509-532.

Novak, M. and Aboul-Ella, F., 1978a. Impedance functions of piles in layered media. Journal of the Engineering Mechanics Division, ASCE, Vol. 104, No. EM6, pp. 643-661.

Novak, M. and Aboul-Ella, F., 1978b. Stiffness and damping of piles in layered media. Proceedings of Earthquake Engineering and Soil Dynamics, ASCE Specialty Conference, Pasadena, California, pp. 704-719.

Novak, M. and El Hifnawy, L., 1983. Effect of Soil-Structure Interaction on Damping of Structures. Earthquake engineering and structural dynamics, Vol. 11, 595-621, 1983.

Vickery, B.J., 1995. Shell/Liner Interaction on a Flexible Foundation. Presented at CICIND Conference, Paris, France.

Wong, H. L. and Luco, J. E., 1985. Tables of impedance functions for square foundations on layered media. International Journal of Soil Dynamics and Earthquake Engineering, Vol. 4, No. 2, pp. 64-81.

Seismic Behavior of Tall Buildings Supported on Pile Foundations

Yingcai Han & **Derek Cathro**
EBA Engineering Consultants Ltd., Edmonton, Canada

Abstract

Seismic behavior of tall buildings supported on piles is examined with respect to linear and nonlinear soil-pile systems, and is compared with the behavior of buildings with rigid base and supported on shallow foundations. The boundary zone model with non-reflective interface is used to account for the nonlinearity of pile foundations, and the validity of the approach calculating foundation stiffness is verified by dynamic experiments on full-scale pile foundations.

Introduction

Many high-rise buildings supported on pile foundations are constructed in soft soil in seismically active areas. The behavior of such structures can be greatly affected by non-linear soil-pile foundation interaction during strong earthquakes. The evaluation of soil-pile-structure interaction is needed in order to establish the forces expected to act on the structure and the piles in a seismic event.

A theoretical analysis of pile-structure interaction is often used for design. Adequate for routine design is a simple procedure based on substructuring and the following assumptions: the input ground motion is given for the level of pile heads and is not affected by the presence of the piles and their cap; soil-pile interaction analysis is conducted separately to yield the pile foundation impedances; and the seismic response is obtained in time domain with input of earthquake records or in frequency domain with input of response spectra. This type of analysis is known as inertial interaction analysis (Novak, 1991).

In this study, the soil-pile system is simulated by a boundary zone model with non-reflective interface. The model is an approximate but simple and realistic method that accounts for the nonlinearity of a soil-pile system. The validity of the model is verified with dynamic experiments on full-scale pile foundations for both linear and nonlinear vibrations. The nonlinear features of the pile foundation and the group effect, known as pile-soil-pile interaction, are examined to provide stiffness and damping of piles in linear and nonlinear cases.

The seismic analysis of a 20-storey building supported on a pile foundation is conducted for different conditions: (1) with rigid base, i.e. no deformation in the foundation; (2) with parameters for a linear soil-pile system; and (3) with parameters for a nonlinear soil-pile system. The effects of pile foundation displacements on the behavior of tall building are investigated and are compared with the behavior of buildings supported on shallow foundation. The effect of using different pile group arrangement on providing a safe and economic pile foundation design is also considered.

Soil-Pile-Structure Interaction

The governing equations of a soil-pile-structure system can be given in the standard form,

$$[m]\{\ddot{u}\} + [c]\{\dot{u}\} + [k]\{u\} = \{P(t)\} \tag{1}$$

in which [m], [c] and [k] are the mass, damping and stiffness matrices incorporating the structure and foundation properties; {u} and {P(t)} are the displacement vector and load vector, respectively.

The number of degrees of freedom can be reduced if the structure is modeled as a shear building. For such a building, the rotations of all masses are the same as the foundation and the number of degrees of freedom is (n+2) for n storey building (Novak and El-Hifnawy, 1983).

With the notation

$[m]$ = diagonal mass matrix listing floor mass $m_1, m_2, \ldots m_n$

$$\{m\} = \left[m_1, m_2, \ldots m_n \right]^T$$

$$\{mh\} = \left[m_1 h_1, m_2 h_2, \ldots m_n h_n \right]^T$$

$$\{u\} = \left[u_1, u_2, \ldots u_n \right]^T$$

and the total mass moment of inertia

$$I = \sum_{1}^{n+1} I_i$$

the governing equation of a shear building can be written as

$$
\begin{bmatrix}
[m] & \{m\} & \{mh\} \\
\{m\}^T & m_b + \sum_1^n m_i & \sum_1^n m_i h_i \\
\{mh\}^T & \sum_1^n m_i h_i & I + \sum_1^n m_i h_i^2
\end{bmatrix}
\begin{Bmatrix} \ddot{u} \\ \ddot{u}_b \\ \ddot{\varphi} \end{Bmatrix}
+
\begin{bmatrix}
[c] & \{0\} & \{0\} \\
\{0\}^T & c_{xx} & c_{x\varphi} \\
\{0\}^T & c_{\varphi x} & c_{\phi\varphi}
\end{bmatrix}
\begin{Bmatrix} \dot{u} \\ \dot{u}_b \\ \dot{\varphi} \end{Bmatrix}
$$

$$
+
\begin{bmatrix}
[k] & \{0\} & \{0\} \\
\{0\}^T & k_{xx} & k_{x\varphi} \\
\{0\}^T & k_{\varphi x} & k_{\varphi\varphi}
\end{bmatrix}
\begin{Bmatrix} \{u\} \\ u_b \\ \varphi \end{Bmatrix}
= -
\begin{bmatrix} \{m\} & m_b + \sum_1^n m_i & \sum_1^n m_i h_i \end{bmatrix}^T \ddot{u}_g(t) \qquad (2)
$$

in which $\ddot{u}_g(t)$ is horizontal ground acceleration, h_i is the height of ith floor. The matrices [k] and [c] list all the stiffness and damping constants of the structure and {0} is the null vector.

The foundation properties are incorporated as shown in equation (2); m_b is mass of foundation; u_b and φ are horizontal translation and rotation; k_{xx}, $k_{x\varphi}$, $k_{\varphi x}$ and $k_{\varphi\varphi}$ are stiffness coefficients of the foundation and c_{xx}, $c_{x\varphi}$, $c_{\varphi x}$ and $c_{\varphi\varphi}$ are damping coefficients of the foundation.

For the analysis of soil-pile-structure interaction, the more difficult but important component is the foundation portion, since there are some uncertain factors in the soil-pile system. By means of the substructure method, the stiffness and damping constants of the pile foundation can be defined separately and then introduced into the governing equations of the structure. Once the impedances of the pile foundation are determined, the procedure of seismic analysis is the same as that for the structure rested on shallow foundation.

Stiffness and Damping of Pile Foundation

The determination of stiffness and damping of piles would serve as a key to obtaining the seismic response of structures supported on pile foundations, but the difficulty is how to evaluate the soil-pile interaction in a nonlinear situation. A number of approaches are available to account for dynamic soil-pile interaction but they are usually based on the assumption that the soil behavior is governed by the law of linear elasticity or visco-elasticity and the soil is perfectly bonded to a pile. In practice, however, the bonding between the soil and the pile is rarely perfect and slippage or even separation often occurs in the contact area. Furthermore, the soil region immediately adjacent to the pile can undergo a large degree of straining, which would cause the soil-pile system to behavior in a nonlinear manner. Both theoretical and experimental studies have shown that the dynamic response of the piles is very sensitive to the properties of the soil in the vicinity of the pile shaft (El-Marsafawi et al.,1992; Han,1989; Han and Navak,1988).

A rigorous approach to the nonlinearity of soil-pile system is extremely difficult and therefore approximate theories have to be used. Novak and Sheta (1980) proposed including a cylindrical annulus of softer soil (an inner weakened zone or so-called boundary zone) around the pile in a plane strain analysis. One of the simplifications involved in the original boundary zone concept was that the mass of the inner zone was neglected to avoid the wave reflections from the interface between the inner boundary zone and the outer zone. To overcome this problem, Veletsos and Dotson (1988) proposed a scheme that can account for the mass of the boundary zone. Some of the effects of the boundary zone mass were investigated by Novak and Han (1990) who found that a homogeneous boundary zone with a non-zero mass yields undulation impedances due to wave reflections from the fictitious interface between the two media, the near field and the far field.

The ideal model of boundary zone should have properties smoothly approaching those of the outer zone to alleviate wave reflections from the interface. Consequently, such a model of boundary zone with non-reflective interface was proposed by Han and Sabin (1995). The presented model of non-reflective interface assumed that the boundary zone has a non-zero mass and a smooth variation into the outer zone by introducing a parabolic variation function, which may be best fit with use of experimental data. Dynamic investigations of piles indicated that the boundary zone model is available to both granular and cohesive soils.

With the impedances of the soil layer, the element stiffness matrix of soil-pile system can be formed in the same way as the general finite element method. Then the overall stiffness matrix of a single pile can be assembled for different modes of vibration. The dynamic stiffness and damping of a single pile can be described in terms of complex stiffness, such as for vertical vibration

$$k_v = \frac{E_p A}{r_o} f_{v1} + i\omega \frac{E_p A}{V_s} f_{v2} \qquad (3)$$

in which r_o is the pile radius; E_p is the Young's modulus of the pile; A is the area of the pile element; ω is the circular frequency; $i = \sqrt{-1}$; and V_s is the shear wave velocity of the soil. f_{v1} and f_{v2} are the dimensionless stiffness and damping parameters.

The validity of the model with non-reflective interface is verified with dynamic experiments on full-scale pile foundations for both linear and nonlinear vibrations. The field tests of pile group were carried out at the Institute of Engineering Mechanics, Harbin, China. The subsurface investigation indicated that the test site was underlain by relatively homogeneous layers of silty clay with occasional lenses of sandy clay down to a depth of 30 meters. The ground-water table was established at 20 meters below the ground surface.

The pile group was comprised of six cast-in-place reinforced concrete piles; each pile was 7.5 m long and 0.32 m in diameter. The pile's slenderness ratio (L/d) is 23.4 and spacing ratio (s/d) is 2.81, where L is the pile length, s is the pile spacing, and d is the pile diameter. The shear wave velocity and mass density of the soil were measured for different layers, such as V_s = 130 m for the top layer and V_s = 280 m at depth of 8 m. Details of the soil were presented in Han and Novak (1992). Two exciters were used in the tests, a smaller one was used to produce linear vibrations of the pile group and a larger one was used for nonlinear vibrations.

Linear Vibration of Pile Group

A smaller exciter, weighing 1.18 KN, was fixed on the pile cap by foundation bolts to produce the harmonic excitation. Several levels of excitation intensities were used in the experiments, as identified, θ = 8, 14 and 28 correspond to excitation intensities of 96, 171 and 259 kg.mm, respectively. The maximum horizontal amplitude of the pile group was measured to be 0.1 mm at top of the pile cap; this is considered to be a small amplitude vibration or say linear vibration.

Under horizontal excitation, the cap produced coupled horizontal and rocking vibration. For linear vibration, the amplitude response of the group can be normalized by

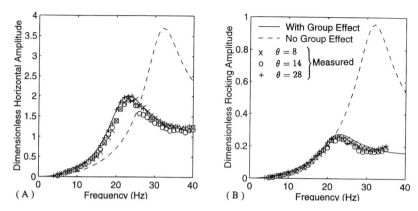

Fig. 1 Linear vibration of pile group under harmonic excitation (A) horizontal response; and (B) rocking response.

the excitation intensity to give dimensionless amplitudes. A comparison of the experimental results with the theoretical predictions is shown in Fig.1 for horizontal and rocking vibration of the pile cap. The boundary zone model with non-reflective interface is employed and the theoretical predictions were done in two ways. First, the pile-soil-pile interaction is accounted for, and the calculated curves are shown by solid lines in Fig.1. The relevant parameters of the boundary zone are as follows: boundary zone thickness ratio, $t_m/r_o = 0.5$; shear moduli ratio, $G_i/G_o = 0.1$; damping ratio in boundary zone, $\beta_i = 0.07$; damping ratio of outer zone, $\beta_o = 0.035$; and Poisson's ratio of soil, $\nu = 0.3$. Second, the group effect is ignored, without the pile-soil-pile interaction, and the calculated results are shown by the dashed lines. It can be seen that the theoretical predictions with the pile-soil-pile interaction match the measured data quite well, but the theoretical solution without the group effect would result in a higher resonant frequency and larger amplitude.

The group effect is considered using the interaction factor approach. The dynamic interaction factors were presented in a chart form by Kaynia and Kausel (1982). The experimental results indicate that the stiffness of group reduced and damping increased due to the group effect.

Nonlinear Vibration of Pile Group

A larger exciter, weighing 4.9 kN, was fixed on the pile cap, and the active component of the horizontal excitation was situated 0.2 m above the cap surface. As

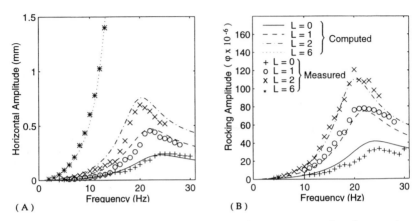

Fig. 2 Nonlinear vibration of pile group under harmonic excitation (A) horizontal response; and (B) rocking response.

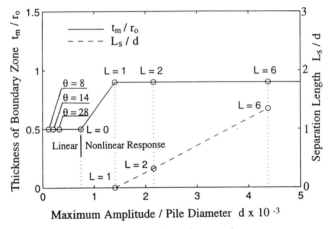

Fig. 3 Effect of excitation intensity on parameters of boundary zone.

indicated, L = 0,1,2 and 6 correspond to the excitation intensities of 472,887,1360 and 3870 kg.mm, respectively. The maximum horizontal amplitude was measured to be 1.4 mm at top of the pile cap, with a corresponding maximum acceleration of 1.13g. This represents a rather intense harmonic vibration, resulting in a nonlinear vibration of the pile group as described in the following paragraphs.

The theoretical predictions and measurements of horizontal and rocking displacements on the cap of the group in different excitation intensities are shown in

Fig.2. It should be explained that the results of rocking vibration for L = 6 were missing because of problems encountered in measuring the rocking response for this case. From the response curves, it can be observed that the resonant frequencies reduce with increasing excitation intensity and the peak amplitudes are not proportional to the excitation intensity at all frequencies. These are typical features of nonlinear vibration.

The boundary zone model gives an insight into the soil-pile interaction. This model provides for the gradual expansion of the yield zone as the excitation level increases. In matching the measured data, allowance had to be made for the modification of boundary zone parameters and the pile separation. To illustrate the influence of excitation intensity on the boundary zone, the variation of parameters in the zone is shown in Fig.3. The maximum amplitude on the cap top was normalized by pile diameter to represent the excitation intensity. It can be seen that the parameters of boundary zone remained in constant for excitations of θ = 8,14,28 and L = 0. In this stage, it was referred as linear vibration. When the excitation intensity increased to larger than L = 0, the thickness ratio, t_m/r_o, increased from 0.5 r_o to 0.9 r_o. When the excitation increased further to larger than L = 2, inferred separation occurred and the separation length, L_s/d, increased with the excitation intensity. This range was referred as nonlinear vibration, caused by the separation effect and the soil softened in boundary zone.

Some of the salient features of the nonlinear vibration of the pile group are listed in Table 1. It can be seen that the stiffness of pile group reduces with increasing excitation intensity. For instance, the horizontal stiffness reduced by almost half when the excitation intensity increased from L = 0 to L = 6. This reduction can principally be attributed to an increase in the thickness of the boundary zone and the soil-pile separation effects.

Table 1. Behavior of the Pile Group in Nonlinear Vibration

Excitation intensity L	Resonant frequency (Hz)	Horizontal		Rocking	
		Stiffness (MN/m)	Damping ratio,D_x	Stiffness $(MN.m)$	Damping ratio,D_ψ
0	24	114	0.34	540	0.039
1	22	97.7	0.29	497	0.039
2	20	87.7	0.25	467	0.039
6	15.8	54.0	0.20	431	0.018

It can be noted that the theoretical predictions agree with the measured data quite well for both horizontal and rocking vibrations. It can be concluded that the employed mathematical model, incorporating a boundary zone, is capable of capturing the nonlinear behavior of a pile group. The results show that the resonant frequency of the pile group reduced and the resonant amplitude increased as the excitation intensity increased. For instance, the resonant frequency reduced from 24 Hz to 15.8 Hz when the excitation level increased from L = 0 to L = 6.

Seismic Response of Tall Building

A 20-storey building is examined in this study as a typical structure supported on a pile foundation in soft soil in a seismically active area. The tall building was

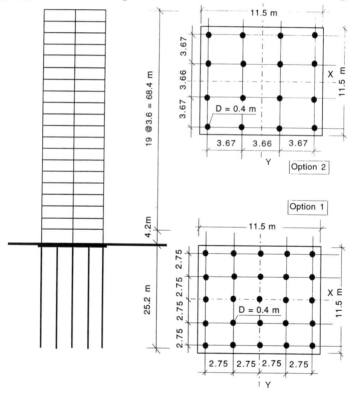

(A) 20-story building supported with pile foundation

(B) Arrangement of piles

Fig. 4. Layout of 20-storey building and two options of pile arrangement.

modeled as a shear building, thus allowing the seismic analysis to be based on equation (2). The seismic response of the building was analyzed using the substructure method, which allows the superstructure and foundation portions to be analyzed separately. The pile impedances were determined as discussed in the preceding section.

The consequence of soil-pile-structure interaction is illustrated by comparing the results of different foundation options. The 20-storey building is 72.6 m tall, with a square cross section and two spans as shown in Fig. 4 (A). The length and width of the building are 11.5 m, with width to height ratio of greater than 1:6. The total dead and live load is 7.0 kN/m², and the weight for each floor is 925 kN. The story stiffness is 0.714 x 10⁶ kN/m.

The soil properties are varied with depth and characterized by the shear wave velocity and unit weight, as shown in Table 2. The soil in the site is considered to be soft soil.

Table 2. Soil Properties in the Site

Depth (m)	Unit Weight (t/m^3)	Shear Wave Velocity (m/s)
0-2	1.90	90
2-4	1.80	90
4-7	1.95	150
7-12	1.95	180
12-15	1.80	210
15-22	1.90	210
22-24	2.00	240
24-40	2.20	460

The piles are precast concrete with a diameter of 0.4 m, installed by driving to depth of 24 m. There are two options for the pile arrangement, as shown in Fig. 4 (B); option 1 is 5 x 5, 25 piles and option 2 is 4 x 4, 16 piles. The seismic analysis is done based on the option 1 at first. The cap of piles is a concrete block with thickness of 1.2 m and weight of 3,700 kN.

The Wilson-θ method of step-by-step integration with respect to time (Clough and Penzien, 1975) was used to calculate the seismic response of the 20-storey building supported on the pile foundation, where θ = 1.37 and time step is 0.02 second. The time history of the horizontal ground acceleration employed is shown in Fig. 5 (A), with the peak value of 0.11 g, from the record of San Fernando earthquake.

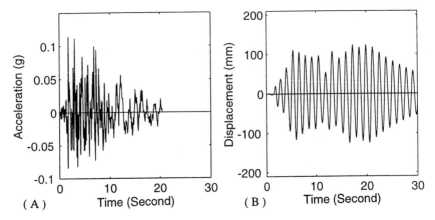

Fig. 5. (A) Time history of ground acceleration from San Fernando earthquake; and (B) seismic response on the top of 20th floor of building with rigid base.

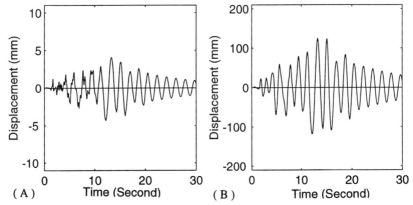

Fig. 6. Seismic response of building with linear soil-pile system; (A) on foundation; and (B) on top of 20th floor.

The seismic analysis for the tall building is conducted for three different foundation conditions to investigate the influence of foundation displacements on the superstructure: (1) with rigid base; (2) with parameters for a linear soil-pile system; and (3) with parameters for a nonlinear soil-pile system.

For the case with rigid base, the stiffness of the foundation is assumed to be infinite and no deformation occurs in the footing. Initial seismic analyses were done

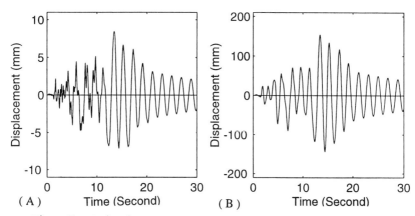

Fig. 7. Seismic response of building with nonlinear soil-pile system; (A) on foundation; and (B) on top of 20th floor.

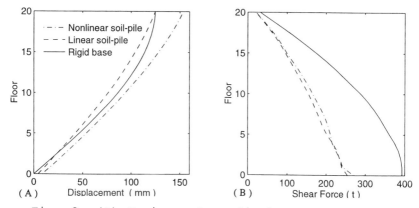

Fig. 8. (A) Maximum story displacements; and (B) maximum interstory shear of building with different conditions of foundation.

in this way forty years ago, when soil-structure interaction was not considered. The time history of displacements at the top of 20th floor are calculated and shown in Fig. 5 (B).

For the case of linear soil-pile system, the stiffness and damping for the pile foundation are calculated based on homogeneous soil layers, without the boundary zone. With such impedance of pile foundation incorporated in eq.(2), the seismic response at the foundation and at the top of 20th floor are calculated and shown in Fig.6.

The nonlinearity of the pile foundation is accounted for by means of the boundary zone model. In this case, the parameters of boundary zone are selected as: $G_i/G_o = 0.1$, $t_m/r_o = 1.0$, $\beta_i = 0.1$ and $\beta_o = 0.05$. With the impedance of pile foundation in nonlinear case, the seismic response at the foundation and at the top of 20th floor are calculated and shown in Fig. 7.

The maximum story displacements and interstory shear of the building in different conditions of foundation are computed and shown in Fig. 8. The natural frequencies, maximum base shear and overturning moment of the building in different foundation conditions are listed in Table 3, in which ω_1, ω_2 and ω_3 are the first three natural circular frequencies of the structure.

Table 3. Natural Frequency, Base Shear and Overturning Moment
in Different Footing Conditions

Condition of Footing	Rigid Base	Linear Case	Nonlinear Case
ω_1	4.77	3.41	3.27
ω_2	14.30	13.50	12.87
ω_3	23.75	22.67	21.50
Max. Base Shear (t)	392.5	254.9	270.5
Max. Overturning Moment $(t-m)$	18400	10080	10800

In the seismic analysis, the internal (structural) damping is taken as $c_i = 0.02 \times k_i / \omega_1$, where k_i is the story stiffness of ith floor. Fig. 5 to Fig. 8 and Table 3 illustrate the seismic behavior of the 20-storey building in different foundation conditions. It can be seen that in the case of rigid base the natural frequencies are higher, and the maximum interstory shear, base shear and overturning moment are quite larger than those in other cases. The results based on the rigid base do not represent the real seismic response, since the soil-structure interaction is ignored. When the flexibility of foundation is accounted for in the case of linear soil-pile system, the natural frequencies are reduced and the seismic response is buffered, such that the maximum interstory shear, base shear and overturning moment are reduced remarkably. When the nonlinearity of pile foundation is accounted for, the natural frequencies are reduced further. The story displacements are computed to be larger than those in the linear case, with a particularly distinct increase in the seismic response at the foundation.

To examine the effect of pile-soil-pile interaction, the dynamic calculation is repeated with no group effect. The stiffness and damping of the group computed in this way are much different from those with the group effect, as described in next paragraph. The group effect is best illustrated by the group efficiency ratio, GER, which is defined as the group stiffness (or damping) divided by the sum of the stiffness (or damping) of all single piles in the group. Since the natural frequencies of the tall build are low as shown in Table 3, the static interaction factors are used to calculate the group effect (El-Sharnouby and Novak, 1985).

In the option 1 pile arrangement, the pile spacing is 2.75 m. The horizontal stiffness of $0.289 \times 10^6 KN/m$ with group effect and $0.812 \times 10^6 KN/m$ without group effect induce a GER of 0.36. In option 2, the pile spacing is 3.67 m. The horizontal stiffnesses are $0.254 \times 10^6 KN/m$ with group effect and $0.517 \times 10^6 KN/m$ without group effect, giving a GER of 0.49. Comparing the two options, the total number of piles is reduced for option 2 but the group efficiency ratio is increased. This results in a final group stiffness reduction of only about 10 percent. The pile-structure analysis indicated that the seismic response of the building with piles of option 2 is similar to that of the building with option 1. Thus, a safe and economic pile foundation design can be expected by selecting a reasonable pile arrangement option.

To compare the effect of a pile foundation against that of a shallow foundation on the seismic behavior of the building, the pile foundation was replaced by a mat foundation having the same size as the cap of piles. The seismic response at the foundation and at the top of 20th

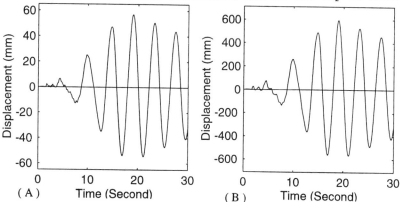

Fig. 9. Seismic response of 20-storey building with a mat foundation; (A) on foundation; and (B) on top of 20th floor.

floor are shown in Fig.9.

From Fig. 9 it can be seen that the seismic response of the building supported on the mat foundation is much different from that on the pile foundation. The first natural circular frequency reduced to $\omega_1 = 1.50$ and the displacements of both the structure and the foundation increased. The different seismic response mainly attributed to changes in the rocking stiffness of the foundation, since the rocking stiffness of the pile foundation is about three times of that of the mat foundation in this case.

Conclusions

The examination by this paper of different seismic behaviors for a tall building supported by different foundations suggests the following conclusions:

1. The seismic behavior of a tall building supported on a pile foundation is different from that on a shallow foundation or a rigid base. Shallow foundations usually yield lower natural frequencies and much larger displacement amplitudes to both the superstructure and the foundation. The theoretical prediction for tall buildings fixed on a rigid base with no soil-structure interaction does not represent the real seismic response, since the stiffness is overestimated and the damping is underestimated.

2. The problem of soil-pile-structure interaction in a seismic environment is complex. The more difficult but important component is the foundation portion. Theoretical and experimental studies on the foundation and superstructure portions can be separated by substructuring, which is a simplified and realistic method.

3. The nonlinear response of a pile foundation can be accounted for approximately by means of the boundary zone model with non-reflective interface. The validity of the model is verified with dynamic experiments on full-size pile foundations for both linear and nonlinear vibrations. When the nonlinearity of piles is accounted for, the natural frequencies of the building are reduced and displacements are increased relative to the linear case.

4. The nonlinearity of pile foundations and the group effect of pile-soil-pile interaction are two important factors for the seismic response of tall buildings. A reasonable seismic analysis for tall buildings supported

on pile foundations is needed to produce a safe and economic design.

Acknowledgments

The financial support by EBA Engineering Consultants Ltd. and the Natural Science and Engineering Research Council of Canada is greatly appreciated.

References

Clough,R.W.and Penzien,J.(1975)."Dynamics of structures". McGraw-Hill Inc.

El-Sharnouby, B. and Novak, M. (1985)."Static and low frequency response of pile groups." Canadian Geotech. J., Vol.22,No.1, 79-94.

El-Marsafawi, H., Han, Y.C. and Novak, M. (1992). "Dynamic experiments on two pile groups." J. Geotech. Eng., ASCE, 118(4), 576-592.

Han, Y.C. and Novak, M. (1988)."Dynamic behavior of single piles under strong harmonic excitation." Canadian Geotech. J., vol.25, No.3, 523-534.

Han, Y.C. (1989)."Coupled vibration of embedded foundation." J. Geotech. Eng., ASCE, 115(9),1227-1238.

Han, Y.C. and Novak, M. (1992)."Dynamic behavior of pile group." China Civil Eng. J., Vol.25,No.5, 24-33.

Han, Y.C. and Sabin, G. (1995)."Impedances for radially inhomogeneous viscoelastic soil media." J. Eng. Mech., ASCE, 121(9), 939-947.

Kaynia, A.M. and Kausel, E. (1982)."Dynamic behavior of pile groups." 2nd Int. Conf. on Num. Methods in Offshore Piling, Austin,TX, 509-532.

Novak, M. and Sheta, M. (1980)."Approximate approach to contact problems of piles." Proc. Geotech. Eng. Dvi., ASCE National Convention "Dynamic Response of Pile Foundations: Analysis Aspects." Florida, Oct.30,53-79.

Novak, M and El-Hifnawy, L. (1983)."Effect of soil-structure interaction on damping of structures." Earthq. Eng. & Struc. Dyn., Vol.11, 595-621.

Novak, M. and Han, Y.C. (1990)."Impedances of soil layer with boundary zone." J. Geotech. Eng.,ASCE, 116(6), 1008-1014.

Novak, M. (1991)."Piles under dynamic loads: State of the Art." Proc. 2nd Int. Conf. on Recent Advances in Geotech. Earthq. Eng. and Soil Dyn., St. Louis, Vol.3, 2433-2456.

Veletsos, A.S. and Dotson, K.W. (1988)."Vertical and torsional vibration of foundation in inhomogeneous media." J. Geotech. Eng.,ASCE, 114(9), 1002-1021.

EFFECT OF TYPE OF FOUNDATION ON PERIOD AND BASE SHEAR RESPONSE OF STRUCTURES

Sanjeev Kumar[1], Member, ASCE and Shamsher Prakash[2], Fellow, ASCE

ABSTRACT

Fundamental natural period of a structure and maximum base shear are the two most important parameters for the dynamic analysis and design of structures. It is understood that the fundamental natural period of a structure supported on soil is higher than the fundamental period of a similar structure, fixed at the base. The effect of type of foundation on the fundamental natural period and base shear response of structures is presented. The results presented indicate that irrespective of the type of foundation, fundamental natural periods and base shear of structures decrease nonlinearly with the soil shear modulus. Structures supported on surface footings have higher fundamental periods compared to structures supported on embedded footings and pile foundations. The results from the present study are compared with the ATC and NEHRP recommendations. It is shown that the provisions given in ATC and NEHRP may be used for structures supported on shallow footings but may not be applicable for structures supported on pile foundations.

INTRODUCTION

Structures are generally supported on soils unless rock is very near the ground surface. The behavior of structures under static or dynamic loads, founded on soils, is different from that of similar structures founded on rock. This difference in the behavior is because of the phenomenon commonly referred to as *Soil-Structure Interaction (SSI)*. Problems involving dynamic loads differ from the corresponding static problems in two important aspects: the time varying nature of the excitation; and the role played by inertia. The response of a structure supported on soil and excited

---

[1]*Senior Engineer, Geotechnology, Inc., 2258 Grissom Drive, St. Louis, Missouri, USA*
[2]*Professor, University of Missouri-Rolla, Rolla, Missouri, USA*

at the base by a dynamic force, estimated on the basis of static soil-structure interaction models or rigid foundation assumption, cannot authentically match actual performance. Therefore, the *Dynamic Soil-Structure Interaction (DSSI)* effect on the seismic response of a structure should not be overlooked (Luan et al., 1995).

The exact analysis considering soil-structure interaction requires that the structure be considered a part of a larger system which includes the foundation and the supporting medium, and that the effect of the spatial variability of the ground motion and the properties of the soils involved be taken into consideration. However, because of the complexity of the problem, performing such a detailed analysis for all types of structures is not always possible in practice. Therefore, approximate methods to incorporate the effects of soil-structure interaction have often been used. One of the approaches involves modifying the free-field ground motion (ground motion in the absence of any structure or foundation) and evaluating the response of the structure to the modified motion of the foundation. Another approach commonly used involves modifying the dynamic properties of the structure, considering it to be rigidly supported, and evaluating the response of the modified structure to the free-field ground motion (Veletsos, 1983). Some codes now include provisions to reflect the effect of soil-structure interaction using the later approximate method. Most of the present codal provisions on soil-structure interaction have mainly developed following the recommendations of Structural Engineers Association of California (SEAOC) and Applied Technology Council (ATC).

Kumar (1996) proposed a simplified model to perform soil-structure interaction analysis of shear type structures with flexible foundations. In the present study, fundamental natural periods are calculated using the model developed by Kumar (1996) for three types of foundations: surface footing; embedded footing; and pile foundation to study the effect of type of foundation on the fundamental period response of shear type structures. The fundamental periods from the simplified model are compared with the recommendations of the ATC 3-06 (1978) and NEHRP (1994). Results are also presented to show the effect of variation in prediction of periods from the two methods on the base shear response of the structures.

SIMPLIFIED MODEL TO PERFORM DYNAMIC SOIL-STRUCTURE INTERACTION

Soil-structure interaction analysis under seismic excitation can be conveniently performed in three steps (Gazetas et al., 1992, Veletsos et al., 1988):

1. Obtain the motion of the foundation in the absence of the superstructure. The resulting motion is called "foundation input motion".

2.   Determine the dynamic impedances of the foundation (spring and dashpot coefficients) associated with translation, rocking, and cross-rocking oscillations.

3.   Compute the seismic response of the superstructure supported on springs and dashpots of Step 2, and subjected at its base to the foundation input motion of Step 1.

For Steps 1 and 2 various formulations have been developed and published in the literature (for Step 1: Idriss and Sun, 1992; Faccioli, 1991; Rosset, 1977; Schnable et al., 1972 and for Step 2: Dobry and Gazetas, 1985, 1988; Gazetas, 1984, 1991; Gazetas et al., 1992; Novak, 1974; Novak and El-Sharnouby, 1983; Roesset and Angelides, 1980). The response of the system, excited by a base motion having acceleration of $\ddot{x}_g$, is governed by equation (1).

$$[M]\{\ddot{x}\} + [C]\{\dot{x}\} + [K]\{x\} = -[M_F]\{1\}\ddot{x}_g \qquad (1)$$

where $[M]$, $[C]$, and $[K]$ are mass, damping, and stiffness matrices, respectively; $\{x\}$ is the column vector of displacements relative to the position of the structure after free-field displacement; $[M_F]$ is the matrix to calculate force at each degree of freedom; a dot superscript denotes differentiation with respect to time, t; and $\{1\}$ is a column vector of ones. It should be noted that $[M_F]$ is differentiated from $[M]$ in equation (1). Generally $[M_F]$ is assumed same as $[M]$. However, it is shown later that this assumption may not be appropriate for all types of problems.

In practice the mass, stiffness, and damping matrices of a combined foundation-structure system are assembled from the corresponding matrices of the two subsystems (foundation and superstructure) as shown in Fig. 1 (Chopra, 1995). The portion of these matrices associated with common degrees-of-freedom at the interface between the two subsystems include contribution from both subsystems. Kumar (1996), and Kumar and Prakash (1997) have shown that the formulation of matrices as shown in Fig. 1 is not appropriate when rotation of the base/foundation is considered. Kumar (1996) developed a simplified model to perform soil-structure analysis of shear type structures with flexible foundations. For an n-story shear type structure, response can be calculated using equation (2).

Notations used in equation (2) are explained in Appendix II. Development of the model is discussed in detail elsewhere (Kumar, 1996, and Kumar and Prakash, 1997). Equation (2) is the simplified equation of motion for an n-story shear type structure, which also includes secondary forces (forces due to P-Δ effect) and time dependent external forces. The form of this equation is similar to equation (1). Vector $\{P_i(t)\}$ on the right hand side of equation (2) is the vector of external forces.

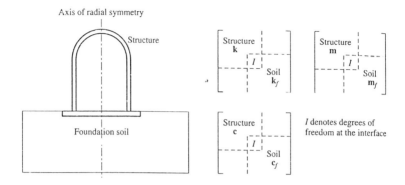

Fig. 1 Assembly of mass, stiffness, and damping matrices from corresponding matrices of subsystems (after Chopra, 1995)

$$
\begin{bmatrix}
m_b & 0 & 0 & 0 & ---- & 0 \\
0 & \rho_b + \sum_{n-1}^{n} \rho_n & 0 & 0 & ---- & 0 \\
0 & 0 & m_1 & 0 & ---- & 0 \\
0 & 0 & 0 & m_2 & ---- & 0 \\
| & | & | & | & & \\
0 & 0 & 0 & 0 & & m_n
\end{bmatrix}
\begin{Bmatrix}
\ddot{x}_b \\
\ddot{\theta}_b \\
\ddot{x}_1 \\
\ddot{x}_2 \\
| \\
\ddot{x}_n
\end{Bmatrix}
$$

$$
+
\begin{bmatrix}
c_x + c_1 & c_1 l_1 + c_{x\phi} & -c_1 & 0 & --- & 0 \\
c_1 l_1 + c_{\phi x} & c_\phi + \sum_{n-1}^{n} c_n l_n^2 & c_2 l_2 - c_1 l_2 & c_3 l_3 - c_2 l_2 & --- & -c_n l_n \\
-c_1 & c_2 l_2 - c_1 l_1 & c_1 + c_2 & -c_2 & --- & 0 \\
0 & c_3 l_3 - c_2 l_2 & -c_2 & c_2 + c_3 & --- & 0 \\
| & | & | & | & & \\
0 & -c_n l_n & 0 & 0 & & c_n
\end{bmatrix}
\begin{Bmatrix}
\dot{x}_b \\
\dot{\theta}_b \\
\dot{x}_1 \\
\dot{x}_2 \\
| \\
\dot{x}_n
\end{Bmatrix}
$$

$$+ \begin{bmatrix} k_x+k_1 & k_1\ell_1+k_{\phi x} & -k_1 & 0 & --- & 0 \\ k_1\ell_1+k_{\phi x}+\overset{n}{\underset{n-1}{\sum}}m_n g & k_\phi+\overset{n}{\underset{n-1}{\sum}}k_n\ell_n^2 & k_2\ell_2-k_1\ell_1-m_1 g & k_3\ell_3-k_2\ell_2-m_2 g & --- & -k_n\ell_n-m_n g \\ -k_1 & k_2\ell_2-k_1\ell_1 & k_1+k_2 & -k_2 & --- & 0 \\ 0 & k_3\ell_3-k_2\ell_2 & -k_2 & k_2+k_3 & --- & 0 \\ | & | & | & | & & \\ 0 & -k_n\ell_n & 0 & 0 & & k_n \end{bmatrix} \begin{Bmatrix} x_b \\ \theta_b \\ x_1 \\ x_2 \\ | \\ x_n \end{Bmatrix}$$

$$= - \begin{bmatrix} m_b & 0 & 0 & 0 & ---- & 0 \\ 0 & 0 & 0 & 0 & ---- & 0 \\ 0 & 0 & m_1 & 0 & ----- & 0 \\ 0 & 0 & 0 & m_2 & ---- & 0 \\ | & | & | & | & & \\ 0 & 0 & 0 & 0 & & m_n \end{bmatrix} \begin{Bmatrix} 1 \\ 1 \\ 1 \\ 1 \\ | \\ 1 \end{Bmatrix} \ddot{x}_g + \begin{Bmatrix} P_{bx}(t) \\ P_{b\theta}(t) \\ P_1(t) \\ P_2(t) \\ | \\ P_n(t) \end{Bmatrix} \qquad (2)$$

Equation (2) shows that formulation of mass stiffness and damping matrices as shown in Fig. 1 is not appropriate when rotation of the foundation is considered. Also, $[M_F]$ is not the same as the $[M]$ for problems involving rotation of the foundation. For problems where the effect of secondary forces (P-Δ effect) is not required, the terms associated with mg should be set to zero. Similarly, if the analysis is required to be performed only for base motion, the external force vector on the right hand side of equation (2) should be set to zero.

The scope of this paper is to study the undamped fundamental natural periods and corresponding base shear response of structures. Therefore, the damping matrix of equation (2) is not used to develop the results presented here. Also, when the foundation soil is assumed to behave as linear elastic, the fundamental natural period of the system depends on the system properties only and, therefore, the terms on the right hand side of equation (2) are not used to compute the fundamental natural periods.

PARAMETERS USED

Structure. Three structures (2-, 4-, and 6-story) are considered in the present study. The structures considered consist of: story height = 3.5 m; size of each column in a story = 300 x 300 mm (2 columns in each story); mass of each story = 19.9 Mg/m^3; and mass moment-of-inertia of each story about its own axis = 38.2 $kN \cdot m \cdot sec^2$. The mass and mass moment-of-inertia of each story are calculated from assumed floor dimensions. Stiffness of the structure are calculated using the procedures given by Chopra (1995).

Foundation. The structures are assumed to be supported on three types of foundation: surface footing; embedded footing; and pile foundation. The shallow footings supporting each column have dimensions of 1.6m X 1.6m X 1.0m. Actual and effective depths of embeddment of embedded footings are 0.8 m and 0.6 m, respectively. Pile foundation for each column consists of four HP 12X72 steel H-piles at a center-to-center spacing of 0.9 meters. Length of each pile is 9 m. Each pile cap is assumed to have dimensions of 1.6m X 1.6m X 1.0m. Pile caps have actual and effective depths of embeddment of 0.8 m and 0.6 m, respectively. Stiffness and damping of foundations is calculated as recommended by Gazetas (1991).

Soil. Soil is assumed to be linear elastic. Modulus is assumed to be constant with depth. However, the lateral and rotational pile stiffnesses have been modified for parabolic soil profile as recommended by Novak and El-Sharnouby (1983) to account for reduction in soil shear modulus near the ground surface. The fundamental natural period of the structures is calculated for a wide range of soil shear modulii (from 15,000 kPa to 170,000 kPa) to model very loose or soft soils to very dense or stiff soils.

FOUNDATION STIFFNESS AND STRUCTURE PERIODS

Figures 2 and 3 show the variation of lateral and rotational foundation stiffness with low strain soil shear modulus, respectively. Both figures show that lateral and rotational stiffnesses of the surface footing are lower than that of embedded footing and pile foundation for all values of soil shear modulus. Lateral stiffness of embedded footing is higher than that of pile foundation but the difference is insignificant for very loose or soft soils. Rotational stiffness for pile foundation is significantly higher than that of embedded foundation. Unlike the lateral foundation stiffness, the difference increases as the soil modulus decreases.

Figure 4 shows the variation of the fundamental natural periods of the system (soil and structure) for the 2-story structure and for all three types of foundations considered in the present study. Fundamental natural period of the 2-story, fixed base structure, is also shown on Fig. 4. This figure shows that, irrespective of the type of foundation, the fundamental natural period corresponding to all three types of foundations decrease nonlinearly with soil shear modulus even though the foundation stiffnesses vary almost linearly with the soil shear modulus. The fundamental period of the flexible based structure approaches the fundamental period of the similar structure, fixed at the base, as the soil modulus increases. In general, the fundamental

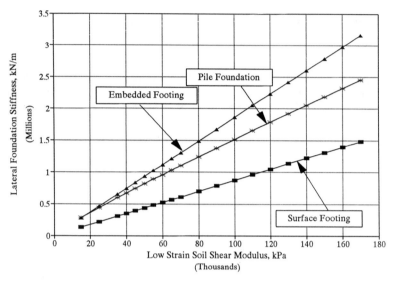

Fig. 2 Lateral foundation stiffness with low strain soil modulus

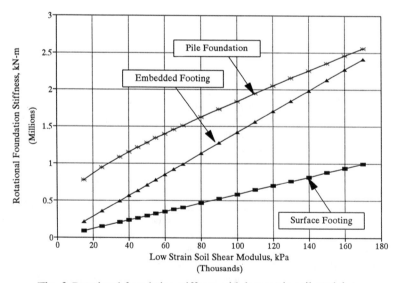

Fig. 3 Rotational foundation stiffness with low strain soil modulus

period response of structures appear to be consistent with lateral foundation stiffness shown in Fig. 2. Structures supported on the surface footings are observed to have highest fundamental periods for all values of soil shear modulus. Fundamental period of structures supported on embedded footing are observed to be slightly higher compared to the fundamental period of structures supported on pile foundations except for very loose or soft soils. Also, the total variation in periods is maximum for structures supported on surface footing (total variation from 0.91 sec @ G_{max} of 15,000 kPa to 0.53 sec @ G_{max} of 170,000 kPa) and minimum for structures supported on pile foundations (total variation from 0.66 sec @ G_{max} of 15,000 kPa to 0.51 sec @ G_{max} of 170,000 kPa). Similar observations are made from the response of the 4- and 6-story structures. Results for 4- and 6-story structures are presented as Figures 5 and 6, respectively.

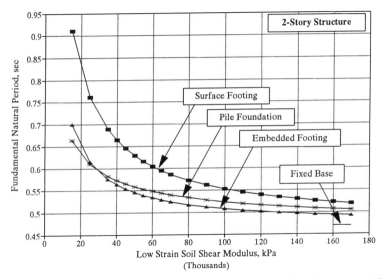

Fig. 4 Variation of fundamental natural periods with low strain soil shear modulus (2-story structure)

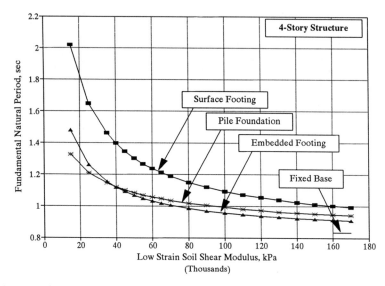

Fig. 5 Variation of fundamental natural periods with low strain soil shear modulus
(4-story structure)

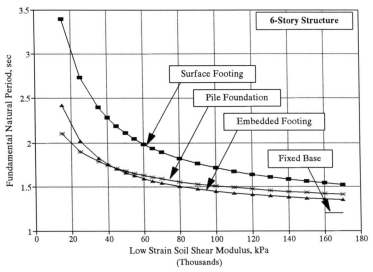

Fig. 6 Variation of fundamental natural periods with low strain soil shear modulus
(6-story structure)

COMPARISON WITH ATC RECOMMENDATIONS

As per the recommendations of the Applied Technology Council, ATC 3-06 (1978) and NEHRP (1994), the fundamental natural period of a structure with flexible base can be computed using equation (3), if the lateral and rotational stiffness of the foundation, and the fundamental natural period of the similar structure, fixed at the base, are known.

$$T' = T \sqrt{1 + \frac{k'}{k_x}\left(1 + \frac{k_x h'^2}{k_\phi}\right)}$$ (3)

where

T'	= Period of a flexible based structure
T	= Period of a structure fixed at the base
h'	= Effective height of the building
k'	= Effective stiffness of the building when fixed at the base
k_x	= Lateral stiffness of the foundation
k_ϕ	= Rotational stiffness of the foundation

For the soil, foundation, and structure parameters used in the present study (presented earlier), periods are computed using equation (3) for the three structures (2-, 4-, and 6-story structures) and the three types of foundations (surface footing, embedded footing, and pile foundation). These periods are compared with the periods computed using the simplified model proposed by Kumar (1996) and Kumar and Prakash (1997). For the structures supported on surface footings, the comparison of fundamental natural periods is presented in Fig. 7. Similar comparisons for the structures supported on embedded footing and the pile foundations are presented in Figures 8 and 9, respectively.

Figure 7 shows that, for the structures supported on surface footings, the ATC recommendations predicted periods close to those predicted using the simplified model. Similar observations are made for the structures supported on embedded footings (Fig. 8). However, for pile foundations (Fig. 9), the fundamental natural periods predicted using the ATC recommendations are significantly different from those predicted using the simplified model. Note that the provisions given in the ATC and NEHRP are primarily developed from the studies of the systems in which foundation is idealized as a shallow rigid mat. However, because of the limited amount of information available concerning the interaction effects for structures supported on spread footings or pile foundations, these provisions are made applicable to the structures founded on spread footings and pile foundations without any substantiation to determine design earthquake forces and corresponding displacements of buildings.   Figure 9 shows that the ATC and NEHRP recommendations for computing periods of structures founded on pile foundations may not be appropriate.

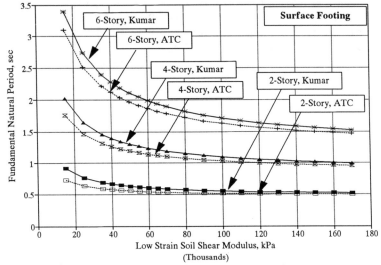

Fig. 7 Comparison of fundamental periods (surface footings)

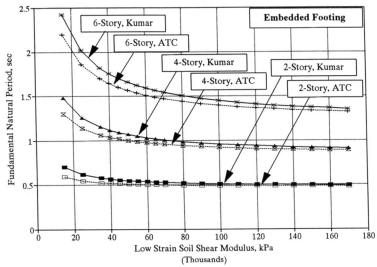

Fig. 8 Comparison of fundamental periods (embedded footings)

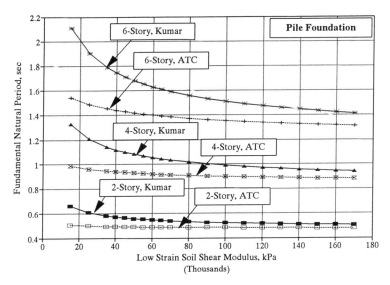

Fig. 9 Comparison of fundamental periods (pile foundation)

The variation in the fundamental periods of structures founded on piles (Fig. 9) has been observed to be more prevalent at lower values of soil modulus (loose or soft soils) compared to the variation at higher soil modulii values (dense or stiff soils). This is particularly significant because piles are used in soft and loose soils which have low soil shear modulus compared to stiff and dense soils. Also, the variation is higher for taller structures.

BASE SHEAR RESPONSE

Depending on the characteristics of the structure, and the ground motion under consideration, soil-structure interaction may increase, decrease, or have no effect on the maximum forces induced in the structure (ATC 3-06). However, based on the recommendations given in the ATC, soil-structure interaction will reduce the design values of base shear from the levels applicable to a fixed base condition. When the structural period and foundation characteristics are known, the calculation of reduction in base shear because of soil-structure interaction effects, as per the ATC recommendations, using the equivalent lateral force procedure, is briefly discussed below. For detailed discussion on the procedure to compute base shear is presented in ATC 3-06.

$$\Delta V = \left[C_s - C_s' \left(\frac{0.05}{\beta'} \right)^{0.4} \right] W'$$ (4)

where

ΔV = reduction in base shear because of soil-structure interaction
C_s = the seismic design coefficient computed using fundamental natural period of fixed base structure
C_s' = the seismic design coefficient computed using fundamental natural period of flexible based structure
β' = the fraction of critical damping for the structure foundation system
W' = effective gravity load of the building

The seismic design coefficient C_s or C_s' can be computed using equation (5) and appropriate fundamental natural period of the structure.

$$C_s \ or \ C_s' = \frac{1.2 A S}{R \left(T_{fix} \ or \ T_{flexible} \right)^{2/3}} \leq \frac{2.5 A}{R}$$ (5)

where

A   = effective peak acceleration at the rock surface
S   = the site coefficient based on the soil profile at a particular site
R   = the response modification factor

Figure 10 shows percent reduction in the base shear computed using equation (4) and fundamental natural periods computed using ATC recommendations and simplified model proposed by Kumar (1996). The effective peak acceleration at rock surface, the site coefficient, and the response modification factor used to develop Fig. 10 are 0.1g, 2.0, and 1.25, respectively. The site coefficient of 2.0 is selected based on a soil amplification study performed by Kumar (1996). As per the ATC recommendations, the value of the response modification factor range between 1.25 and 8.0. Value of 1.25 was selected for ordinary frames.

It is clear from Fig. 10 that for structures supported on pile foundations, the prediction of lower periods using ATC recommendations (Fig. 9) may have significant effect on the base shear response of structures. Reduction in base shear using period predicted by the ATC recommendation is significantly lower than the reduction in base shear computed using the period computed with the simplified model proposed by Kumar (1996). Similar to the fundamental period response, base shear decreases nonlinearly with the soil shear modulus. Percent reduction in the base shear using periods computed from the ATC recommendations is approximately 7% for very loose to very soft soils compared to approximately 23% reduction computed using periods predicted by the simplified model. For very dense and stiff soils, the percent reduction in the base shear in both the cases is less than 5%.

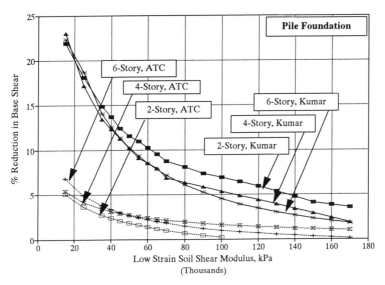

Fig. 10   Base shear response with low strain soil shear modulus (pile foundation)

CONCLUSIONS

The results presented show that irrespective of the type of foundation, fundamental natural periods of structures and percent reduction in base shear because of soil-structure interaction decrease nonlinearly with the soil shear modulus. Structures supported on surface footings have higher fundamental periods compared to structures supported on embedded footings and pile foundations. The ATC and NEHRP recommendations for calculation of fundamental period of structures using the fundamental natural period of fixed base structures may not be applicable for structures supported on pile foundations. Also, reduction in base shear because of soil-structure interaction effects for structures supported on pile foundations, computed using the periods predicted by the ATC recommendations, can be significantly lower than the reduction in base shear computed using the periods predicted by the simplified model proposed by Kumar (1996). The ATC and NEHRP recommendations to compute fundamental natural period of flexible based structures may be used for structures supported on shallow footings resting on the surface of the elastic half space or embedded in to the ground. However, it is the authors' opinion that the ATC recommendations should not be used for pile supported structures to compute fundamental natural period of structures

APPENDIX I. REFERENCES

ATC (1978). "Tentative provisions for the development of seismic regulations for building." ATC 3-06, Applied Technology Council.

Chopra, A. K. (1995). *Dynamics of structures-theory and applications to earthquake engineering*. Prentice Hall, Englewood Cliffs, New Jersey.

Dobry, R., and Gazetas, G. (1985). "Dynamic stiffness and damping of foundations by simple methods." *Vibration Problems in Geotechnical Engineering*, G. Gazetas and E. T. Selig, eds., ASCE Annual Convention, Detroit, Michigan, 75-107.

Dobry, R., and Gazetas, G. (1988). "Simple method for dynamic stiffness and damping of floating pile groups." *Geotechnique*, 38(4), 557-574.

Faccioli, E. (1991). "Seismic amplification in the presence of geologic and topographic irregularities." *Proc. 2nd Int. Conf. on Recent Adv. in Geotech. Earthquake Engrg. and Soil Dyn.*, St. Louis, Missouri, 1779-1797.

Gazetas, G. (1984). "Seismic response of end-bearing single piles." *Soil Dyn. and Earthquake Engrg.*, 3(2), 82-93.

Gazetas, G. (1991). "Foundation vibrations." *Foundation Engineering Handbook*, H.Y. Fang, ed., Van Nostrand Reinhold, 553-593.

Gazetas, G., Fan, K., Tazoh, T., Shimizu, K., Kava, and Markis, N. (1992). "Seismic pile-group-structure interaction." *Piles Under Dynamic Loads, Geotech. Engrg. Div.*, S. Prakash, ed., Geotechnical Special Publication No. 34, ASCE, 56-94.

Idriss, I.M. and Sun, J.I. (1992). "SHAKE91: A computer program for conducting equivalent linear seismic response analysis of horizontally layered soil deposits." University of California-Davis, California.

Kumar, S. (1996). "Dynamic Response of Low-Rise Buildings Subjected to Ground Motion Considering Nonlinear Soil Properties and Frequency-Dependent Foundation Parameters", Ph.D. Thesis, University of Missouri-Rolla, Rolla, Missouri.

Kumar, S. and S. Prakash. (1997) "A Simplified Model to Perform Dynamic Soil-Structure Interaction Analysis", International Journal of Numerical and Analytical Methods in   Geomechanics (submitted).

Luan, M., Lin, G., and Chen, W. F. (1995). "Lumped-parameter model and nonlinear DSSI analysis." *Paper No. 5.02, Proc. 3rd Int. Conf. on Recent Adv. in Geotech. Earthquake Engrg. and Soil Dyn.*, St. Louis, Missouri, 355-360.

NEHRP(1994). "Recommended provisions for the development of seismic regulations for new building." Report No. 222A, Federal Emergency Management Agency.

Novak, M. (1974). "Dynamic stiffness and damping of piles." *Can. Geotech. J.*, 11(4), 574-598.

Novak, M., and El-Sharnouby, B. (1983). "Stiffness and damping constants for single piles." *J. Geotech. Engrg.* ASCE, 109(7), 961-974.

Rosset, J.M. (1977). "Soil amplification of earthquakes." *Numerical methods in Geotechnical Engineering*, edited by Desai, C.S. and Christian, J.T., McGraw-Hill, 639-642.

Rosset, J.M. and Angelides, D. (1980). "Dynamic stiffness of piles" *Numerical methods in offshore piling*, London, England, 75-81.

Schnable, P.B., Lysmer, J., and Seed, H.B. (1972). "SHAKE: a computer program for earthquake response analysis of horizontally layered sites." Report EERC 72-12, University of California, Berkeley.

Veletsos, A.S. (1983). "Seismic design provisions for soil-structure interaction." International Workshop on Soil-Structure Interaction, Roorkee, 3-9

Veletsos, A.S., Prasad, A.M., and Tang, Y. (1988). "Design approaches for soil-structure interaction." Technical report NCEER-88-0031

APPENDIX II. NOTATIONS

The following symbols are used in this paper:

c_i = Damping of the ith story columns

c_x = Damping of the foundation in translation

$c_{x\phi}$ = Damping of the foundation in cross-rotation

c_ϕ = Damping of the foundation in rotation

$c_{\phi x}$ = Damping of the foundation in cross-rotation

g = Acceleration due to gravity

k_i = Stiffness of the ith story columns

k = Constant

k_x = Stiffness of the foundation in translation

$k_{x\phi}$ = Stiffness of the foundation in cross-rotation

k_ϕ = Stiffness of the foundation in rotation

$k_{\phi x}$ = Stiffness of the foundation in cross-rotation

ℓ_i = i^{th} Story height from center-to-center of the masses

m_i = Mass of ith story

m_b = Mass of the foundation

n = Number of stories

x_b = Horizontal displacement at the base (foundation displacement) due to soil-structure interaction only. Note that this does not include free-field displacement

x_g = Free-field displacement of the system

x_i = Total displacement of ith mass with respect to the position of the structure after free-field displacement

$\ddot{x}_g$ = Acceleration of the base motion

θ_b = Rotation of the base

ξ_i = Modal damping ratio for i^{th} mode

ρ_b = Mass moment of inertia of the base about its center of gravity

ρ_i = Mass moment of inertia of the ith story mass about its center of gravity

ω_i = i^{th} Natural frequency of the structure

Soil-Pile-Structure Interaction in Liquefying Sand
from Large-Scale Shaking-Table Tests and Centrifuge Tests

T. Kagawa[1], Y. Taji[2], M. Sato[2] , and C. Minowa[3]

Abstract

An international joint research project is underway to study the dynamic response and failure mechanisms of underground structures, such as pile foundations and buried pipelines, in potentially liquefiable soils. The project is called EDUS (Earthquake Damage to Underground Structures) project, and it is organized and administered by WSU (Wayne State University) and NIED (National Research Institute for Earth Science and Disaster Prevention, Science and Technology Agency of Japan). One of the key tasks of this project is to produce reliable experimental data that can be used to improve currently available numerical techniques and design guidelines for pile foundations and buried pipelines. In this paper, results from large-scale shaking-table tests and centrifuge tests are introduced and compared, and the results are analyzed by numerical methods to gain insight into the dynamic behavior of pile foundations in liquefying sand.

Introduction

A number of pile foundations have been found damaged and failed during historically major earthquakes. For example, case histories of damaged pile foundations in the city of Niigata have been reported by various researchers in recent years [Hamada and O'Rourke, 1992], and an increasingly larger number of reports on damaged pile foundations are becoming available from the last Kobe earthquake in 1995. The key causes of the damage and failures of pile foundations in such earthquakes were liquefaction and lateral ground movements due to

---

[1]Department of Civil and Environmental Engineering, Wayne State University, Detroit, Michigan 48202
[2]Institute of Technology, Shimizu Corporation, 3-4-17 Echujima, Koto-ku, Tokyo, Japan
[3]National Research Institute for Earth Science and Disaster Prevention, Science and Technology Agency, 3-1 Tennoudai, Tsukuba, Ibaraki, Japan

liquefaction.   A number of pile foundations, of course, have been found damaged due to excessive inertial loading from superstructures and large soil deformation due to strong ground shaking.   These damage and failures of piles suggest inadequacy of the current seismic design methods for pile foundations.   Therefore, efforts are needed to improve our understanding of the dynamic behavior of pile foundations and to improve our design analysis methods.

An international joint research project is underway to study the dynamic response and failure mechanisms of underground structures, such as pile foundations and buried pipelines, in potentially liquefiable sandy soils.   The project is called EDUS (Earthquake Damage to Underground Structures) project.   The project is organized and administered by WSU (Wayne State University) and NIED (National Research Institute for Earth Science and Disaster Prevention, Science and Technology Agency of Japan).   One of the key tasks of this project is to produce reliable experimental data that can be used to improve currently available numerical techniques and design guidelines for pile foundations and buried pipelines.   Large-scale shaking-table tests, centrifuge tests, and small-scale model tests are being employed to accomplish the ultimate objective of this project.   In 1996, fabrication of a large-scale laminar shear box (length=12m, height=6m, and width=3.5m) was completed at NIED in Tsukuba, Japan, and an initial series of large-scale shaking-table tests were conducted by using this new shear box.   This was the first time in history that prototype-size steel and pre-stressed concrete piles embedded in loose saturated sand layers were subjected to simulated earthquake shaking.

This paper first highlights the major findings from the large-scale shaking-table tests performed in 1996 and from the initial series of centrifuge tests simulating the large-scale shaking-table test conditions.   The paper then discusses comparisons between the experimental and numerical analysis results to gain insight into the fundamental mechanism of the interaction between the liquefying sand and pile foundations.

Large-Scale Shaking-Table Test

The large-scale shaking-table test was performed at NIED in Tsukuba, Japan.   The size of the shaking table is approximately 15 m by 15 m, and the payload is 500 tons.   The maximum displacement and velocity amplitudes of the driving system are 22 cm and 75 cm/sec.   The shaking table is powerful enough to apply an acceleration of 500 gals at this payload.   On this shaking table, the newly fabricated large-scale laminar shear box was mounted and used for the present study.   The model ground (i.e., the sand layer) and the model pile foundation for the large-scale shaking-table tests are shown schematically in Fig. 1.

Sand Layer was of submerged loose sand.   The test sand for this study was obtained in the city of Hokota near the Lake Kasumigaura in Japan.   The sand was washed at the sampling site to remove excessive fine contents. The test sand was first air-dried and it was placed into the laminar shear box, partially filled with water, using a free-fall method. No compaction efforts were made to densify the sand layer. The height of the sand layer thus prepared was 5.93 m; the average void ratio was 0.81; and the relative density was 38.3 %. The sand layer was completely submerged, without a partially saturated surficial layer. Since the large-scale test reported here was one of the first series in our plan, the test models and conditions were preferably as simple as possible to make subsequent numerical analyses easier. The cases with a non-liquefiable surface layer will be studied in 1997 and after.

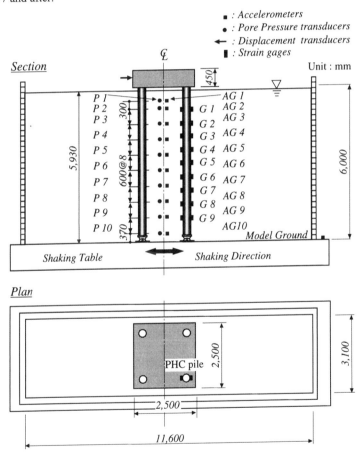

Fig.1 Soil-Pile-Structure Model in Large-Scale Shaking-Table Test

Model Pile Foundation   The model pile foundation consisted of four pre-stressed concrete piles with an outside diameter of 0.3 m, an inside diameter of 0.128 m, and a length of 6 m. The model pile foundation was installed into the laminar shear box before filling the laminar shear box with the test sand. Efforts were made not to densify the sand around the piles to prepare the sand layer in a homogeneous condition as much as possible. The piles were pinned at their tips, but they were grouted into the structural mass at their heads. The spacing ratio (= center-to-center distance/diameter) of the piles was six to minimize pile-group interaction effects. Responses of piles with pile-group interaction effects will be studied in 1997 and after. The pre-stressed concrete piles supported a steel block with a weight of 22.2 tons. The steel block represented the inertial effects of typical mid-rise buildings with a number of stories of about 10.

Measurement System   consisted basically of the strain gages attached to the reinforcing bars in the pre-stressed concrete piles and of accelerometers and pressure transducers in the sand layer and on the piles. The layout of such measurement transducers is shown in Fig. 1. The accelerometers and pressure transducers in the sand layer were secured onto a rolled plastic net, which should have essentially followed the movement of the sand layer during the test. The effectiveness of this installation method was confirmed in our preliminary series of shaking-table tests on model piles conducted in 1992 through 1995.

Shaking-Table Input Motion   was the NS component of the ground acceleration recorded at a depth of 32 m at a site on the Port Island in the city of Kobe during the last Hyogoken Nambu earthquake in 1995. At the recording site, significant liquefaction occurred in the silty sand and gravel layers at depths less than 32 m. Therefore, this motion was able to liquefy the sand layer and to bring the piles into the failure conditions. The recorded motion has a peak acceleration of approximately 560 gals, but the amplitudes of this motion was scaled down to have a peak acceleration of 350 gals. After this test, the same motion was input to the shaking table without scaling to bring the piles into total failure conditions. After the test, the insides of the piles were visually inspected by lowering a video camera into the piles. The test sand was then removed from the shear box, and the failure conditions of the piles were inspected and recorded. Along the entire lengths of the piles, a number of transverse cracks were seen, with deeper cracks concentrated mainly in the top two thirds of the piles. Results of this test will be reported elsewhere.

Centrifuge Tests

Introduction Centrifuge testing methods intend to reproduce the stress-strain conditions in a model that are identical to those in the corresponding prototype by artificially increasing the apparent gravitational acceleration using a centrifugal force. A number of researchers have used the methods to study the fundamental mechanism of liquefaction and soil-structure interaction problems [Arulanadan and Scott, 1993].

The main objective of running centrifuge tests in the EDUS project was to compare results from the large-scale and centrifuge tests to generate ample experimental data to verify and improve centrifuge modeling techniques for pile foundations and buried pipelines. Therefore, although the centrifuge test was performed after the large-scale shaking-table test was conducted, measured responses of the sand layer and the piles were not made available to the researchers engaged in the centrifuge test. The centrifuge test was intended to be a blind prediction of large-scale test results. The only information provided to the researchers for the test included: 1) the geometry and dimensions of the sand layer and the model pile foundation, the method of preparation of the sand layer, and 2) the recorded acceleration of the shaking table in the large-scale test.

The centrifuge facility used in our study is detailed by Sato [1994]. The shaking table inside the centrifuge bucket is driven by an electromagnetic device. The shaker exhibits stable high-frequency performance even in the centrifuge environment. However, like any other electromagnetic shakers, the shaker lacks in power in the low-frequency range. This becomes critical especially when the geometrical scaling ratio λ is small and the frequencies of interest are relatively low.

		Symbol	Scale ratio	Unit	1g	Centrifuge
Sand Stratum	Thickness	H_g	$1/\lambda$	m	5.93	0.374 (5.61)
	Density	ρ_t	1	g/cm^3	1.83	1.98
Pile	Length of pile	L	$1/\lambda$	m	6.0	0.4
	Diameter	D	$1/\lambda$	mm	300	20
	Thickness	t	$1/\lambda$	mm	86	0.5 (5.7)
	Young's modulus	E	1	MN/m^2	44,100	206,000 (44,100)
	Geometrical moment of inertia	I	$1/\lambda^4$	cm^4	36,300	0.146 (0.717)
	Bending stiffness	$E \cdot I$	$1/\lambda^4$	MN·m^2	16.0	0.00030 (0.00032)
	Area	A	$1/\lambda^2$	cm^2	471	0.306 (2.09)
	Normal stiffness	$E \cdot A$	$1/\lambda^2$	MN	2,080	6.30 (9.22)
Footing	Mass	m_f	$1/\lambda^3$	kg	22,200	7.1 (6.6)
	Length	L_f	$1/\lambda$	m	2.5	0.16
Exciting acceleration		α	λ	g	0.2	3.1

$1/\lambda$ = model / prototype = 1 / 15

Table 1 Similitude Relationships in Centrifuge Test

Similitude Relations and Model The similitude relationships employed in the present centrifuge test are summarized in Table 1. The geometrical dimensions of the large-scale model were reduced down by a factor of fifteen. Therefore, the apparent gravitational acceleration was increased to fifteen g's, and time was reduced by fifteen times. The experimental model in our centrifuge test is shown in Fig. 2. The model is essentially geometrically similar to its prototype. The sand layer was prepared using the same test sand and the method employed in the large-scale test.

As indicated in Fig. 2, the response measurement system employed in the centrifuge test was similar to its prototype (i.e., strain gages on the pile and accelerometers and pressure transducers in the sand and on the pile). However, the number of transducers that could be installed was naturally less than that in the large-scale test due to space limitations.

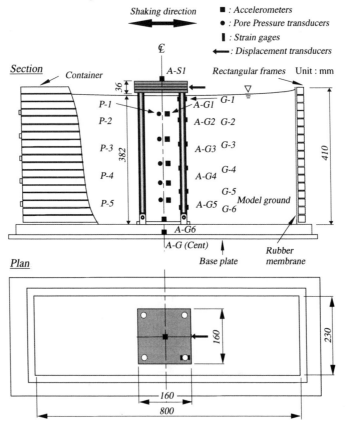

Fig.2 Soil-Pile-Structure Model in Centrifuge Test

Large-Scale versus Centrifuge Test Results

Introduction Figures 3 through 5 compare time histories of the acceleration, excess pore-water pressures in the sand layer, and the bending moment in the piles from the large-scale and centrifuge tests, respectively. Figures 6 and 7 compare profiles from the large-scale and centrifuge tests of the excess pore-water pressures and of the bending moment in the pile.

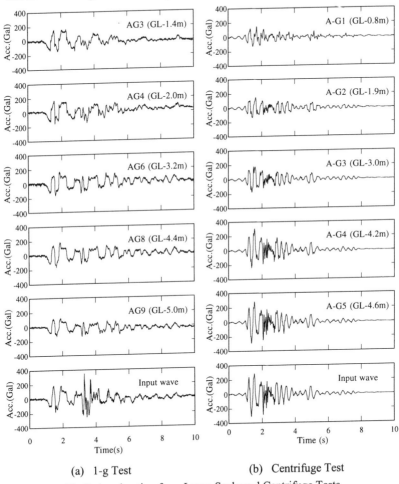

(a) 1-g Test   (b) Centrifuge Test

Fig.3 Acceleration from Large-Scale and Centrifuge Tests

Figure 3 clearly indicates that the shaking-table motion (i.e., "Input wave") in the centrifuge test is different from that in the large-scale test. The table motion in the centrifuge test is dominated by the frequency components that are higher than those in the large-scale test.   This difference was imposed by the limited low-frequency performance of the shaker in the centrifuge bucket, and it resulted in significant difference in the responses of the sand layer and the piles in the large-scale and centrifuge tests.   However, this was unavoidable in our tests in 1996.   Based on this lesson, in the next series of tests in 1997, we plan to use, in our large-scale tests, the shaking-table motions recorded in the centrifuge tests. This will provide us with opportunities for making better comparisons of results from these two tests.

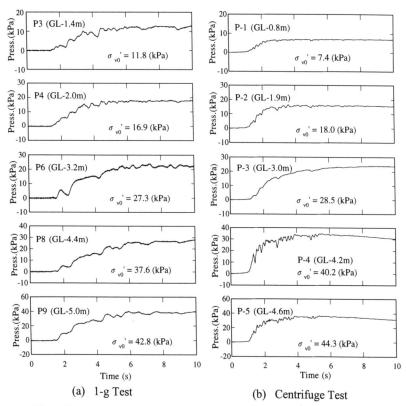

(a) 1-g Test   (b) Centrifuge Test

Fig.4 Excess Pore-Water Pressures from Large-Scale and Centrifuge Tests

Figure 3 indicates that: 1) the acceleration responses of the sand layer in the centrifuge test contain higher frequency components than those in the large-scale test; and 2) the peak acceleration is less in the centrifuge test, compared to the large-scale test, towards the top of the sand layer and is higher near the bottom.

This difference should be attributed to the difference in excess pore-water buildup and dissipation in the two tests. In the centrifuge test, the sand layer was excited more severely than in the large-scale test especially in the early part of shaking, and excess pore-water pressures developed more rapidly in the centrifuge test. Also, excess pore-water pressures dissipated quicker in the centrifuge test. These are shown in Figs. 4 and 6. These figures also show that in the large-scale test liquefaction advanced from the top and the bottom of the sand layer and that no liquefaction occurred in the middle part of the sand layer. This general trend is also seen in the centrifuge test.

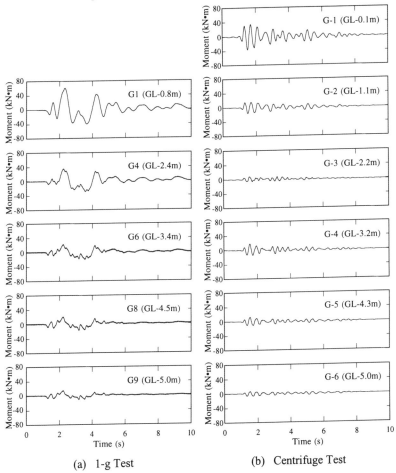

(a) 1-g Test   (b) Centrifuge Test

Fig.5 Bending Moment in Large-Scale and Centrifuge Tests

The shaking-table motion in the centrifuge test had higher frequency components compared to those in the large-scale test, and excess pore-water pressures developed more rapidly in the centrifuge test. These differences caused difference in observed pile moments. Figure 5 shows that the moment in the pile in the large-scale test is larger in the upper part of the pile. The pile moment in the

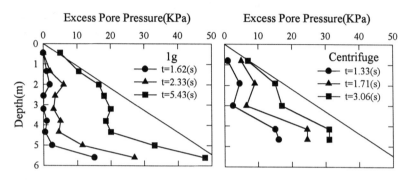

Fig.6 Excess Pore Pressures from Large-Scale and Centrifuge Tests

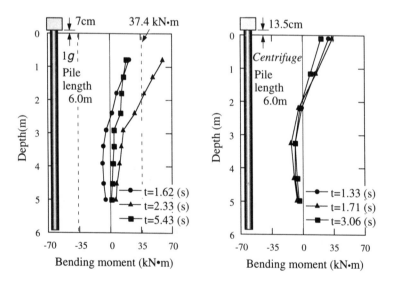

Fig.7 Bending Moment from Large-Scale and Centrifuge Tests

centrifuge test does not show the same trend: i.e., the largest moment is at G-1; the peak moment decreases at G-3; it again increases at around G-4 and G-5; and it then decreases. Also, the pile moment is shown to be smaller in the centrifuge test.

These differences are due to the difference in the vibrational modes of the piles in the two tests. Figure 7 indicates that the pile in the large-scale test vibrated at the second mode in the early stage of shaking (t=1.62 sec.) and then at the first mode thereafter (t=2.33 and 5.43 sec.). However, the pile in the centrifuge test vibrated at the second mode from the beginning to the end of the test (t=1.31, 1.71, and 3.06 sec.). Therefore, the pile moment in the centrifuge test was generally smaller than that in the large-scale test, except near the pile head. Incidentally, the threshold moment above which cracking may occur in the prototype pile was 37.4 kN-m.

Numerical Analyses by SRANG and NONSPS

Comparisons between the large-scale and centrifuge tests were not satisfactory for the tests conducted in 1996 and reported here. Therefore, results from the large-scale test were studied to evaluate the numerical analysis methods, SRANG and NONSPS [Kagawa, 1996], and to obtain better understanding of the response behavior of the large-scale model to prepare for the experiments in 1997 and after.

Dynamic Responses of the Sand Layer were computed by SRANG. The computer program SRANG can perform an effective-stress based, one-dimensional site-response analysis in the time domain. Various stress-strain models including elasto-plastic constitutive models are incorporated into the program. In this study, however, a simple stress-strain model similar to the hyperbolic and Ramberg-Osgood models was used. The stress-strain relations of soil were supplied to SRANG in the form of small-strain shear moduli (G_{max}'s) and modulus

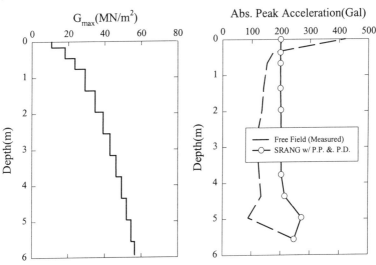

Fig.8 Estimated Small-Strain Moduli   Fig.9 Computed and Measured Acc.

reduction curves (G/G_{max} versus shear strain). These data were needed to define the backbone curves of the stress-strain relations of soil layers. The Masing rule was used to represent unloading and reloading portions of the stress-strain relations.

The small-strain shear moduli (G_{max}'s) were estimated by the equation proposed by Iwasaki, et. al. [Iwasaki and Takagi, 1978] with the estimated average void ratio of the sand layer of 0.81, Fig. 8. As indicated in Fig. 8, the sand layer was divided into eleven sublayers in our analysis to be compatible with the locations of the accelerometers and pressure transducers. The estimated shear moduli in Fig. 8 are generally consistent with the average shear moduli estimated from the measured shear-wave velocity of the sand layer of 135 m/sec. for a slightly denser state of the sand layer with a void ratio of 0.77. The modulus reduction curves were estimated for each sublayer using the empirical equations proposed by Ishibashi and Zhang [1993]. It should be noted that the modulus reduction curves were considered to be dependent on effective confining stress in our analysis. In addition to the stress-strain data of the test sand, SRANG requires parameter values for excess pore pressure generation [Kagawa and Kraft] and consolidation. Parameter values for the pore pressure generation model were determined to be consistent with the cyclic triaxial test results on the test sand. With SRANG, dynamic responses of the sand layer were computed on an effective-stress basis, where the sand stiffness was altered according to the magnitude of excess pore-water pressures in sand. No dilatancy effects during and after liquefaction, however, were included in the present study. The unit weight and the shearing resistance of the sand layer in our SRANG analysis were adjusted to incorporate the effects on the responses of the sand layer of the weight and the sliding friction of the laminar shear box. These effects, however, were found to be acceptably negligible.

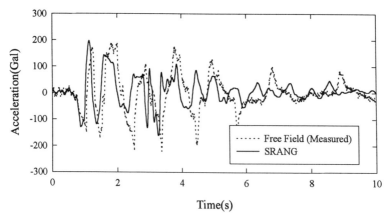

Fig.10 Computed and Measured Acceleration of the Sand Layer

Figure 9 compares measured and computed peak accelerations in the sand layer. The peak acceleration in sand is generally less than that of the shaking-table motion. The large measured peak acceleration near the surface of the sand layer should be discarded because the accelerometer at this location was exposed to air in the early stage of shaking due to the settlement of the sand layer. Figure 10 shows a favorable comparison between the acceleration histories of measured and computed responses near the surface of the sand layer. Results in Figs. 9 and 10 indicate general validity of the one-dimensional site-response technique currently available.

Dynamic Responses of the Piles were computed by NONSPS, and results were compared with measured responses. The computer program NONSPS obtains dynamic responses of a single pile that is laterally supported by a series of soil-pile elements. The soil-pile elements are basically nonlinear springs and dashpots that connect the pile to the free-field soil, and they can represent the following dynamic p-y relations [Kagawa, 1996]:

$$p = E_s \delta y + C \dot{y} \qquad (1)$$

where p = the dynamic lateral soil reaction to a unit length of the pile, E_s = the average Young's modulus of the soil around the pile, which is continually updated to reflect the change in soil stiffness in accordance with the strains induced in soil and with the magnitude of excess pore-water pressures in the soil around the pile, δ = the average soil-pile reaction coefficient, which is approximately equal to unity, y = the pile deflection relative to the free-field soil movement, and C = the dashpot constant. For dynamic soil-pile interaction problems dominated by linearly visco-elastic properties of soils, the dashpot constant may be assumed to be $2\rho(V_S + V_P)D$, where ρ is the mass density of soil, V_S and V_P are the average shear-wave velocity and constrained-wave velocity of the soil, and D is the effective width of the pile. For problems involving liquefaction, however, the dashpot can represent the viscous behavior of liquefied soil. In the present analysis, the dashpot constant was initially set to $2\rho(V_S + V_P)D$, but the constant was successively modified as excess pore-water pressures built up in sand around the pile. The following criterion was used to make this modification:

$$C = (1-u)^{1/4} 2\rho(V_S + V_P)D + 2\rho V_L D u \qquad (2)$$

where u is the average excess pore-water pressure ratio in sand around the pile and V_L is to represent the viscous reaction of liquefied sand. V_L was estimated to be 50 m/sec. based on the results of our previous small-scale shaking-table tests. The average excess pore-water pressure at a depth in sand around the pile was estimated from the pore pressure model [Kagawa and Kraft]. For this estimation, the shear strain in the sand around the pile was estimated from $(1 + \nu)y/(\bar{f}D)$ where ν is the Poisson's ration of the sand and $\bar{f}$ is the average influence factor. In our present analysis, $\bar{f}$ was assumed to be 5.0.

Figure 11 compares the peak acceleration profiles in the piles. The measured and computed peak accelerations in the piles are shown to be higher than those in the sand layer. Figure 11 includes the results from two NONSPS runs. The first run was made by first running SRANG to compute the responses of the sand layer to obtain the free-field soil motion for NONSPS. In this first run, no pore pressure model was used in either SRANG or NONSPS. Therefore, no degradation of sand stiffness occurred during shaking, resulting in high peak accelerations in the piles. In the second run, however, NONSPS was run with; a) the effects of excess pore-water pressure buildup around the piles, and b) the measured acceleration in the sand layer. The computed peak accelerations are slightly lower than the measured ones, but they are in good agreement with measured values.

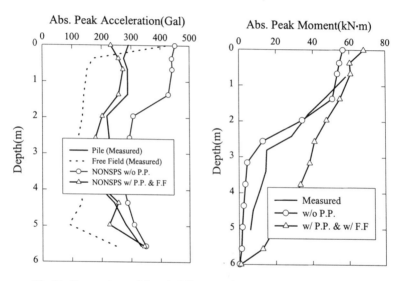

Fig.11 Comparison of Acc. in Pile   Fig.12 Comparison of Pile Moment

Figure 12 compares measured and computed moment in the piles. The first NONSPS run, which does not consider the effects of excess pore-water pressure buildup in the sand layer, tends to underestimate the pile moment, especially towards the bottom section of the pile. On the other hand, the second NONSPS run, which fully includes the effects of excess pore-water pressure buildup in the sand layer around the pile, overestimates pile moment. The overestimation is due to underestimation of the lateral reaction from the sand. In the large-scale test, the mid-height portion of the sand layer did not liquefy as shown in Fig. 6, but our analysis by NONSPS resulted in complete liquefaction of the sand around the pile over the full length of the pile at an early stage of shaking. This suggests that computed pile moment is relatively sensitive to the details of the dynamic p-y relations used. Also, our analysis by NONSPS suggests that: 1) the peak acceleration in pile could be comfortably estimated by neglecting the effects of pore pressure buildup in sand; 2) the pile moment estimated by neglecting the effects of pore pressure buildup tends to underestimate the measured values; and 3) the pile moment obtained from an analysis with the effects of pore pressure buildup in sand around the pile overestimates the measured values. These results indicate the importance of correct evaluation of the lateral reaction from liquefied sand to obtain reliable and rational estimation of pile responses.

Concluding Comments

As part of the joint research efforts for the EDUS project, we conducted an initial series of large-scale and centrifuge tests on prototype-sized pile foundations in liquefying sand layer in 1996. Our main objectives were to: 1) produce reliable experimental data that could be used to calibrate currently available numerical methods and design guidelines for pile foundations, and 2) evaluate and advance dynamic centrifuge testing techniques for such structures by comparing the test results from large-scale and centrifuge tests.

The first objective of the EDUS project was reasonably satisfied for the test models employed in 1996. The second objective, however, was not accomplished in 1996 due mainly to the difference in the performance of the shaking tables in the large-scale and centrifuge tests. However, we expect that the difference can be overcome in our joint research activities for the EDUS project planned in 1997 and after.

Acknowledgments

A number of individuals have contributed to the EDUS project. The authors would like to express their sincere appreciation to Dr. N. Ogawa, NIED, Mr. A. Abe, Tokyo Soil Research Co., Ltd., Mr. T. Sakai, Kisojiban Consultants, Ltd., and Dr. K. Ishihara, Science University of Tokyo. Without their help and encouragement, the EDUS project was not made possible.

The numerical analyses by SRANG and NONSPS were performed by Ms. Xian Tao, a graduate research assistant at Wayne State University.

References

Arulanandan, K. and Scott, R.F., eds. (1993), *Verification of Numerical Procedures for the Analysis of Soil Liquefaction Problems*, Proceedings, Rotterdam.

Hamada, M. and O'Rourke, T.D. (1992), *Case Studies of Liquefaction and Lifeline Performance during Past Earthquakes*, Technical Report NCEER-92-0001.

Ishibashi, I. and Zhang, X. (1993), "Unified Dynamic Shear Moduli and Damping Ratios of Sand and Clay," Soils and Foundations, Vol.33, No.1, 182-191.

Iwasaki, T., Tatsuoka, F., and Takagi, Y. (1978), "Shear Moduli of Sands under Cyclic Torsional Shear Loading," Soils and Foundations, Vol.20, No.1, 45-59.

Kagawa, T. and Kraft, L.M., Jr., "Modeling the Liquefaction Process," Journal, Geotechnical Engineering Division, ASCE, Vol.107, No.GT12, 1593-1607.

Kagawa, T. (1996), *SRANG - Site Response Analysis of Nonlinear Ground V.4*, Geotechnical Engineering Group, Wayne State University, Detroit, Michigan 48202, March.

Kagawa, T. (1996), *NONSPS - Nonlinear Dynamic Response Analysis of Soil-Pile-Structure Systems V.4*, Geotechnical Engineering Group, Wayne State University, Detroit, Michigan 48202, March.

Sato, M. (1994), "A New Dynamic Geotechnical Centrifuge and Performance of Shaking Table Tests," *Proceedings, International Conference Centrifuge 94*, Singapore, 157-162.

Earthquake Induced Forces in Piles in Layered Soil Media

Amir M. Kaynia[1]

Abstract

The purpose of this paper is to study the influence of soil layering on the magnitude of forces induced in pile foundations by a seismic disturbance. The structure is modeled as a shear type single-degree-of-freedom oscillator and the foundation is a typical 3x3 pile group embedded in a layered half space. A three dimensional pile-soil-structure model, which rigorously accounts for the pile-soil-pile interaction, is used to calculate the forces in the piles. Two soil profiles representing sites with a soft soil layer over a more stiff half-space are considered. Three sets of structural parameters, covering practical range of structural properties are used and the analysis is performed for several excitation frequencies. The presented results cover the distribution of shear forces and bending moments in two of the piles in the group. The results are used to examine the influence of stiffness contrast of the soil layers and the structural properties on the distribution of the internal pile forces.

Introduction

Surveys of the performance of piles in recent strong earthquakes have shown that various mechanisms can contribute to the damage or failure of pile foundations. The majority of the reported damages have been attributed to the soil liquefaction and lateral spreading of the soil (e.g., Matsui and Oda (1996) and Tokimatsu et al. (1996)). However, a comprehensive investigation by Mizuno (1987) has concluded that the pile forces arising from the passage of the seismic waves (often referred to as kinematic interaction forces) can also be so large to cause damage. Independent of the mechanism of the damage, these observations point to the need for a reevaluation of the current practice of pile design.

[1] Norwegian Geotechnical Institute, P.O.Box 3930 Ullevaal Hageby, N-0806 Oslo, Norway

One of the commonly used procedures for seismic design of piles is based on a consideration of the inertial interaction forces, while those forces generated during the kinematic interaction phase are assumed negligible (Hadjian et al. 1992). Thus following this procedure, the piles are designed either for the code-specified forces on the structure, or alternatively, the inertia forces in the structure obtained from its dynamic analysis. The numerical results reported by Wolf et al.(1981), Waas and Hartmann (1984) and Masayuki and Shoichi (1991) support such an assumption. For instance, the analysis of a large pile foundation supporting a rigid mass reported by Waas and Hartmann (1984) showed that the pile-head shear forces and bending moments at the soil-structure resonant frequency are much larger than their associated kinematic interaction values.

Considerable research has been conducted in the recent past on the seismic response of pile-soil-structure systems (see Tazoh et al. (1987), Hasegawa et al. (1988), Kobori et al. (1991), Finn and Gohl (1992), Gazetas et al. (1992), Kaynia and Mahzooni (1996) and Makris et al. (1996) among others). Various numerical and experimental methods have been used in these studies and various structural types have been considered. Some of these studies have focused on the behavior of the structure and others have dealt primarily with the piles. These investigations have considerably furthered the understanding of the seismic behavior of pile foundations. However, more has yet to be learned about the fundamental aspects of seismic pile response so that a more reliable seismic design procedure for piles can be produced.

The objective of this paper is to examine the behavior of a representative pile-soil-structure system in a layered half-space. To this end, a 3x3 pile group with typical dimensions supporting a shear-type single-degree-of-freedom (SDOF) structure with various structural properties is considered. In order to investigate the effect of contrasting soil layers on the internal pile forces, two soil profiles, representing sites with surficial soft soil, are considered. The presented results cover the foundation and structural responses as well as the distribution of the internal pile forces under vertically propagating seismic shear waves.

Foundation-Structure Parameters

Figure 1 shows the foundation-structure system under consideration. The foundation is a 3x3 pile group embedded in a two-layered half-space. The soil profile consists of a surficial soft layer of thickness l_1 over a half-space. The soil medium is characterized by its elastic modulus, E_s, mass density, ρ_s, Poisson's ratio, ν_s, and hysteretic damping ratio, ξ_s. A commonly used soil parameter is the shear wave velocity, V_s, which is related to the previously defined soil parameters.

The piles are similarly represented by the same material properties and denoted E_p, ρ_p, ν_p and ξ_p, respectively. In addition, d and l are used to represent the pile's cross sectional diameter and length, respectively and s denotes the center-to-center pile spacing.

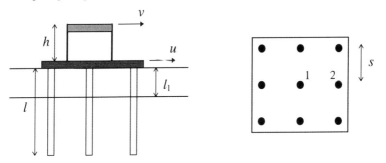

Figure 1. Pile-Soil-Structure System

The structure is modeled as a shear type SDOF oscillator characterized by its mass, m_0, mass moment of inertia, I_0, lateral stiffness, K_0, damping ratio, ξ_0, and height of the mass above ground level, h. The fixed-base natural period of the structure is denoted by T and is given by $2.\pi.(m_0/K_0)^{0.5}$. The mass and mass moment of inertia of the pile cap are taken as zero.

The following values were used for the various system parameters:

- Soil parameters:
  $\rho_s = 1800$ kg/m^3
  $v_s = 0.4$
  $\xi_s = 0.05$
  $V_s = 90$ m/s (top layer)
  $V_s = 200$ m/s (half-space)
  $l_1 = 3.0$ m (Profile 1)
  $\quad = 6.0$ m (Profile 2)

- Structural parameters:
  $m_0 = 300,000.0$ kg
  $I_0 = 0.0$
  $h = 5.0$ m
  $T = 0.2, 0.5, 1.0$ and 1.5 s.
  $\xi_0 = 0.02$

- Pile parameters:
  $E_p = 20$ GPa ($E_p/E_s = 100$ and 500 for halfspace and top layer)
  $d = 0.75$ m
  $l = 15.0$ m ($l/d = 20$)
  $s = 3.75$ m ($s/d = 5$)
  $\rho_p = 2400$ kg/m^3
  $v_p = 0.2$
  $\xi_p = 0.00$

An enhanced version of the numerical code PILES (Kaynia 1982) was used to model the dynamic response of the pile-soil-structure system. According to this formulation the traction that develop at the piles-soil interface as a result of wave passage and inertial interaction are related to the soil motions through the soil Green's function, and to the pile motions through the beam's dynamic stiffness relations. Imposition of the displacement compatibility between the soil and piles, together with the introduction of appropriate equilibrium conditions of the soil-foundation system, leads to the calculation of the motions of the piles and structure as well as the internal pile forces. (The reader is referred to Kaynia and Kausel (1991) for details.)

The present analyses are carried out under steady-state harmonic vibration in the frequency domain. Therefore, all response quantities have complex values. The seismic excitation is assumed to be due to vertically incident shear waves with a unit acceleration (1.0 m/s^2) at the ground surface. The resulting accelerations of the foundation and structure are denoted by u and v, respectively. The other quantities of interest in this study are the absolute values of the shear forces and bending moments along the piles. The values of these quantities, together with their associated kinematic interaction result, are plotted for three excitation frequencies f = 0.5, 1.0 and 2.0 Hz.

Discussion of Results

For a given soil profile and excitation frequency the numerical results are presented for a set of structural parameters corresponding to the periods T = 0.2, 0.5, 1.0, and 1.5s as well as for the no-structure case, referred to as the *kinematic* interaction case. For each set, the results are presented separately for the two soil profiles described in the previous section.

Figure 2 shows the distribution of the maximum earthquake-induced shear force and bending moment along Pile 2 (Fig. 1) in soil profile 1. The excitation frequency is f = 0.5 Hz. The figure shows the results for the above four structural cases as well as their associated kinematic interaction values. A number of important observations emerge from these plots. Perhaps the most important is the relatively large localized forces appearing at about the depth where the stiffness of the soil profile changes abruptly. As is evident in Fig. 2, these forces are strongly contributed by the kinematic interaction of piles. This has important implications on the seismic design of pile in sites with sharp contrasts in layer properties. It is also worth noting that the kinematic shear force is small at the pile-head whereas the bending moment can be quite large. The second observation concerns the strong sensitivity of the pile forces to the characteristics of the structure. The pile forces increase as frequency of excitation approaches the natural frequency of the foundation-structure system. This behavior was expounded by Kaynia and Mahzooni (1996) who also showed how the intensity of forces at around this frequency relates to the total damping of the system.

Figure 3 shows the same set of results for the pile group in soil Profile 2. The plots in this figure show essentially the same features displayed in Fig. 2 except that, as a result of the increase in the thickness of the soft layer, the location of the intensified

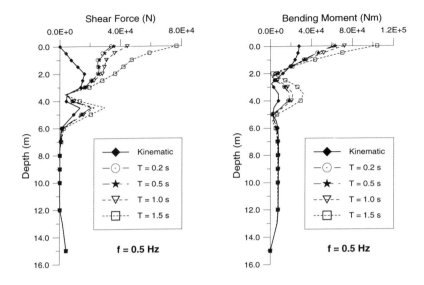

Figure 2. Maximum Seismic Forces along Pile 2 in Profile 1: $f = 0.5$ Hz

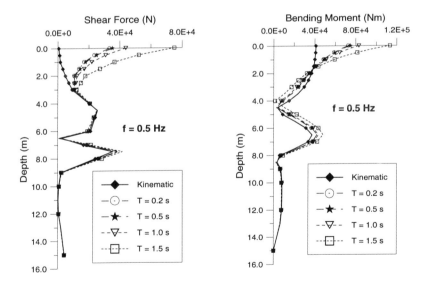

Figure 3. Maximum Seismic Forces along Pile 2 in Profile 2: $f = 0.5$ Hz

pile forces has shifted down. This figure even more clearly illustrates the dominant contribution of the kinematic interaction forces at this depth. It is also interesting to note that, except in a frequency band around the natural frequency of the foundation-structure system, the pile forces contributed by the kinematic interaction can be as large as the total seismic forces. This observation again points to the importance of incorporating the kinematic interaction forces in the seismic design of piles in layered media.

For an excitation frequency $f = 1.0$ Hz the maximum shear forces and bending moments along Pile 2 are shown in Fig. 4 for Profile 1 and in Fig. 5 for Profile 2. The plots in these figures essentially corroborate the conclusions drawn above. It is to be noted, however, that a different set of structural parameters which result in a system with the natural frequency matching the excitation frequency (the structure with $T = 1.0$ s in these figures) gives the largest response

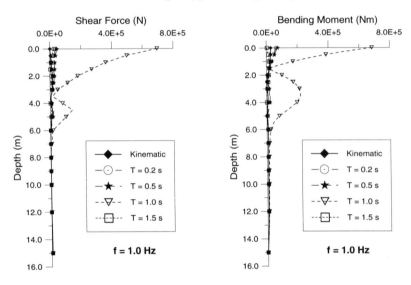

Figure 4. Maximum Seismic Forces along Pile 2 in Profile 1: $f = 1.0$ Hz

Finally, Figures 6 and 7 portray the maximum seismic forces in Pile 2 in the two soil profiles for the excitation frequency $f = 2.0$ Hz. These figures further confirm the earlier conclusions in this section. As regards the group effect, the kinematic seismic forces are almost identical in the piles of a group, whereas the forces arising from the inertial interaction can vary largely in the piles. To clarify this point, the distributions of the maximum pile forces in the center pile (Pile 1) in Profile 2, for excitation frequency $f = 2.0$ Hz are presented in Fig. 8. Comparison

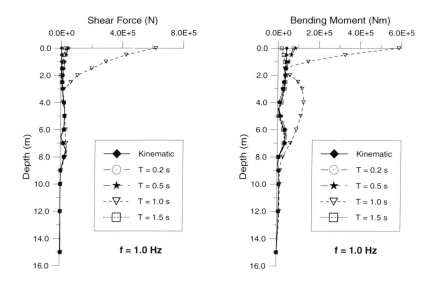

Figure 5. Maximum Seismic Forces along Pile 2 in Profile 2: f = 1.0 Hz

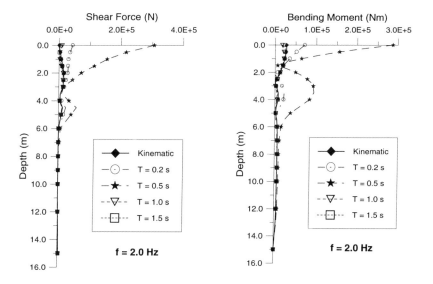

Figure 6. Maximum Seismic Forces along Pile 2 in Profile 1: f = 2.0 Hz

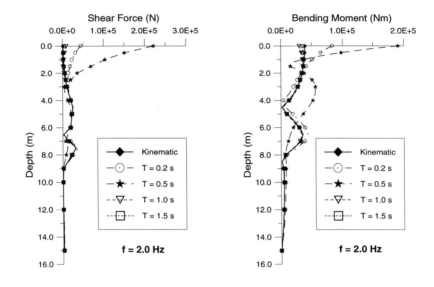

Figure 7. Maximum Seismic Forces along Pile 2 in Profile 2: $f = 2.0$ Hz

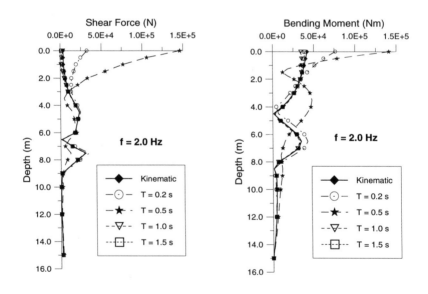

Figure 8. Maximum Seismic Forces along Pile 1 in Profile 2: $f = 2.0$ Hz

between the plots in this figure and those in Fig. 7 (for Pile 2) reveal that the inertial interaction forces are on the average 30% lower in the center pile. The trends displayed by the force distributions are, however, similar in the various piles of the group.

As for the superstructure response, Fig. 9 shows the variations with excitation frequency of the maximum accelerations of the pile cap, u, and the maximum accelerations of the structure, v, in Profile 1. (Similar results were obtained for Profile 2.) As expected, the maximum structure's acceleration occurs at the same frequency where maximum pile forces were observed. Magnification of acceleration by up to 20 are displayed by the structure. Although a typical earthquake, being characterized by a wide frequency content, does not usually induce such large accelerations, the above results nevertheless point to the potential of dramatic seismic responses.

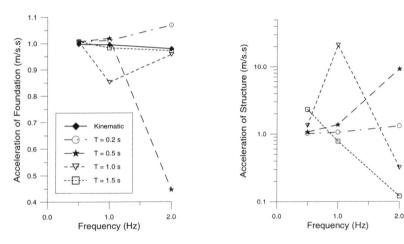

Figure 9. Maximum Acceleration of Pile Cap and Structure in Profile 1

Conclusions

The paper presented a set of typical results for the seismic performance of a 3x3 pile-group foundation supporting a shear-type SDOF structure. A range of structural parameters and two soil profiles exhibiting abrupt changes in the stiffness of the layers were used. The results showed that in a frequency band around the natural frequency of the foundation-structure system, the inertial interaction dominates the structural response and the internal forces of the piles. Outside this frequency band, kinematic interaction can play a significant role in the magnitude of the pile forces at around the layer interfaces. Depending on the system parameters and the stiffness contrast of the layers such forces can be as large as the total forces in the piles.

References

Finn, W.D.Liam and Gohl,W.B. (1992)."Response of model pile groups to strong shaking." *Piles under Dynamic Loads*, Geotech. Special Publ. No. 34, ASCE, S. Prakash, ed., 27-55.

Gazetas,G., Fan,K., Tazoh,T., Shimizu,K., Kavvadas,M. and Makris,N. (1992). "Seismic pile group-structure interaction." *Piles under Dynamic Loads*, Geotech. Special Publ. No. 34, ASCE, S. Prakash, ed., 56-93.

Hadjian,A.H., Fallgren,R.B. and Tufenkjian, M.R. (1992)." Dynamic soil-pile-structure interaction- The state-of-practice." *Piles under Dynamic Loads*, Geotech. Special Publ. No.34, ASCE, S. Prakash, ed., 1-26.

Hasegawa,N., Oshima,Y., Hadjian,A.H., Lau,L. and Fallgren,R.B. (1988). "Modeling for nonlinear response prediction of pile-supported structures." *Proc. 9th World Conf. Earthquake Engrg.*, Tokyo-Kyoto, Japan, 5, 363-368.

Kaynia,A.M. (1982)."Dynamic stiffness and seismic response of pile groups." *Research Report R82-03*, Dept. Civil Engrg., M.I.T., Cambridge, Massachusetts.

Kaynia,A.M. and Kausel,E. (1991)."Dynamics of piles and pile groups in layered soil media." *Soil Dyn. Earthquake Engrg.*, 10(8), 386-401.

Kaynia,A.M. and Mahzooni,S. (1996)."Forces in pile foundations under seismic loading." *J. Engrg. Mech.*, ASCE, 122(1), 46-53.

Kobori,T., Nakazawa,M., Hijikata,K., Kobayashi,Y., Miura,K., Miyamoto,Y. and Moroi,T. (1991)."Study on dynamic characteristics of a pile group foundation." *Proc. 2nd Int. Conf. Recent Advances Geotech Earthquake Engrg. Soil Dyn.*, St. Louis, Missouri, 1, 853-860.

Makris,N., Gazetas,G. and Delis,E. (1996). " Dynamic soil-pile-foundation-structure interaction: records and predictions." *Geotechnique*, 46(1), 33-50.

Masayuki,H. and Shoichi,N. (1991)."A study on pile forces of a pile group in layered soil under seismic loading." *Proc. 2nd Int. Conf. Recent Advances Geotech. Earthquake Engrg. Soil Dyn.*, St. Louis, Missouri, 3, 2079-2086.

Matsui,T and Oda,K. (1996). "Foundation damage of structures." Special issue of *Soils and Foundations*, 189-200.

Mizuno,H. (1987)."Pile damage during earthquake in Japan (1923-1983)." *Dynamic Response of Pile Foundations- Experiments, Analysis and Observation*, Geotech. Special Publ. No.11, ASCE, T. Nogami, ed., 53-78.

Tazoh,T., Shimizu,K. and Wakahara,T. (1987)."seismic observations and analysis of grouped piles." *Dynamic Response of Pile Foundations- Experiment, Analysis and Observation*, Geotech. Special Publ. NO.11, ASCE, T. Nogami, ed., 1-20.

Tokimatsu,K., Mizuno,H. and Kakurai,M. (1996). "Building damage associated with geotechnical problems." Special issue of *Soils and Foundations*, 219-234.

Waas,G. and Hartmann,H.G. (1984)."Seismic analysis of pile foundations including pile-soil-pile interaction." *Proc. 8th World Conf. Earthquake Engrg.*, San Francisco, California, 5, 55-62.

Wolf,J.P., von Arx,G.A., de Barros,F.C.P. and Kakubo,M. (1981)."Seismic analysis of the pile foundation of the reactor building of the NPP Angra 2." *Nuclear Engrg. and Design*, North Holland, Amsterdam, 65, 329-341.

Numerical Implementation of a 3-D Nonlinear Seismic S-P-S-I Methodology

By Y.X. Cai[1] A.M. ASCE, P.L. Gould[2] F. ASCE, C.S. Desai[3] F. ASCE

ABSTRACT: In order to more precisely investigate seismic Soil-Pile-Structure-Interaction (S-P-S-I), a three-dimensional finite element subsystem methodology with an advanced plasticity-based constitutive model for soils has been developed. A computerized implementation of the proposed method is presented in this paper. The numerical modelling of soil plasticity and the solution of nonlinear dynamic equations are discussed in detail.

INTRODUCTION

The proposed three-dimensional nonlinear seismic Soil-Pile-Structure-Interaction (S-P-S-I) model is a finite element sub-system model which consists of two subsystems: one is the structure subsystem and the other is pile-foundation subsystem [Figure 1] (Cai, 1995; Cai, Gould, and Desai, 1995). The two subsystems are connected at the junctions between the pile heads of the foundation and the column bases of the structure. The interaction of these two subsystems is transmitted through the motions and the dynamic forces of the pile heads and the column bases. The constitutive law of the soil medium considered in this study is described by the δ^*-version of the Hierarchical Single Surface (HiSS) plasticity based modelling approach for cyclic behavior of soft clays (Wathugala and Desai, 1993; Desai and Wathugala, 1993).

To simultaneously take account of both kinematic and inertial interactions of the structure subsystem and the pile foundation subsystem and solve for the seismic response for the proposed subsystem model, a successive-coupling incremental solution scheme in the time domain is developed as shown in Figure 2, where F_p represents the pile head forces, $V_b^"$ represents the bedrock accelerations, $V_p^"$ represents the pile head accelerations, and $U_g^"$ represents the ground acceleration input for the column bases (Cai, Gould, and Desai, 1995). Such a successive-coupling incremental procedure can

[1]Engineer, Metropolitan Engineering and Surveying, St. Louis, MO 63012, USA
[2]Prof. & Chair., Dept. of Civ. Engrg., Washington Univ., St. Louis, MO 63130, USA
[3]Prof., Dept. of Civ. Engrg. & Engrg. Mech., Univ. of Arizona, Tucson, AZ 85721, USA

be repeated until the entire response history is determined. When the time step Δt is small enough, the continuous response history is well approximated by the discrete

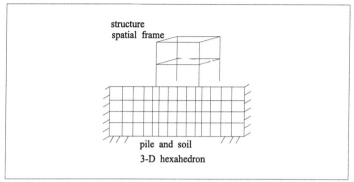

Figure 1.   3-D FE Mesh of Soil-Pile-Structure System.

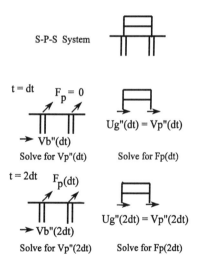

Figure 2.   Schematic of S-C Scheme

step approach. Using this solution scheme, the structure subsystem will be analyzed for the more realistic pile foundation motions, which are different from conventionally used free-field motion.

DYNAMIC EQUATIONS FOR UNEVEN SUPPORT EXCITATIONS

The unique feature of the proposed structure subsystem model is that the structure may be subjected to non-uniform foundation motion (support excitation), which is equivalent to the pile head motion. The non-uniform foundation motion will be obtained from the coupled kinematic and inertial interactions of the soil-pile-structure systems.

To derive the incremental equations of motion for a structure subjected to uneven support excitations, the principle of superposition is assumed to be valid within each incremental time step, provided that the time step is small enough. For a structure subsystem with n active degrees of freedom (DOF) and m support DOF, the dynamic equilibrium of the subsystem due to multiple-support excitations is obtained by the superposition of the dynamic responses of the subsystem due to each independent support input. The incremental equations of dynamic equilibrium of the structure subsystem at time $t+\Delta t$ can be expressed as

$$\left(\frac{4M}{\Delta t^2} + \frac{2C}{\Delta t} + K\right)\Delta U = -MR\ddot{U}_g - K\,^tU + M\left(\frac{4\,^t\dot{U}}{\Delta t} + \,^t\ddot{U}\right) + C\,^t\dot{U}$$

(1)

where Δt is the time step; superscript t indicates the time t; ΔU are increments of the dynamic response vector U, so that $^{t+\Delta t}U = \,^tU + \Delta U$; R is an n x m matrix of the pseudostatic response influence coefficients.

For nonlinear response of the structure subsystem, Equation (1) will be reconstructed by means of the modified Newton-Raphson iteration scheme. The stiffness matrix K and the damping matrix C will be replaced by the corresponding tangent stiffness matrix ^tK and tangent damping matrix ^tC, which will be updated at the beginning of each time step t.

NONLINEAR CONSTITUTIVE LAW OF SOIL

The foundation subsystem, which is composed of piles and the surrounding soils, is idealized as an assemblage of eight-node hexahedral elements. Pile elements may be treated as linear or nonlinear, depending on the refinement of the model. The soils considered in this study are clays similar to a marine clay from Sabine, Texas. This clay has already studied thoroughly by both experimental and numerical methods (Wathugala and Desai, 1993).

The strength of the clay is assumed to be governed by the effective stresses in the soil skeleton, and the strains in the clay are approximated as those of the soil skeleton. Therefore, the finite element formulation for the soil is based on the stress-strain relationship of the soil skeleton. The constitutive law of the soil skeleton is described by an advanced plasticity based model, the δ^*-version of the Hierarchical

Single Surface (HiSS) (Wathugala and Desai, 1990; Wathugala and Desai, 1993; Desai and Wathugala, 1993) modelling approach for cyclic behavior of soft clays.

The associative model of the δ^*-version of the HiSS approach is employed in this study to define the behavior of the soil for different stress-strain regimes such as virgin loading, unloading, and reloading. When a material point is yielding, the stress at this point lies on the yield surface or prestress surface F. At this state, loading is defined as the virgin loading, at all the other states, loading is defined as reloading. The stress-strain relationships for different loading situations are defined as follows:

(1) Virgin Loading
Under the virgin loading, the incremental stress-strain relationship is expressed as

$$d\sigma_{ij} = C_{ijkl}^{VL} \, d\varepsilon_{kl} \tag{2}$$

where

$$C_{ijkl}^{VL} = C_{ijkl}^{e} - \frac{C_{ijnm}^{e} n_{nm}^{F} n_{op}^{F} C_{opkl}^{e}}{H^{VL} + n_{rs}^{F} C_{rstu}^{e} n_{tu}^{F}} \tag{3}$$

where C_{ijkl} = constitutive stiffness tensor. Superscript e denotes elastic quantities and superscript VL denotes quantities associated with virgin loading. The tensor n_{ij}^{F} represents the unit normals to the yield surface F. H^{VL} is the virgin plastic modulus which can be found from the consistency condition in the theory of plasticity as

$$H^{VL} = -\frac{\dfrac{\partial F}{\partial \alpha_{ps}}}{\left(\dfrac{\partial F}{\partial \sigma_{mn}} \dfrac{\partial F}{\partial \sigma_{mn}} \right)^{\frac{1}{2}}} \frac{\partial \alpha_{ps}}{\partial \xi_{v}} \frac{n_{kk}^{F}}{\sqrt{3}} \tag{4}$$

where F = the yield surface of the soil;

$$F = \left(\frac{J_{2D}}{P_{a}^{2}} \right) + \alpha_{ps} \left(\frac{J_{1}}{P_{a}} \right)^{n} - \gamma \left(\frac{J_{1}}{P_{a}} \right)^{2} \tag{5}$$

α_{ps} = the hardening function ;

$$\alpha_{ps} = \frac{h_1}{\xi_v^{h_2}} \tag{6}$$

J_1 = the first invariant of the stress tensor, σ_{ij}; J_{2D} = the second invariant of the deviatoric stress tensor; P_a = the atmospheric pressure; n, γ, h_1, h_2 = material parameters which can be determined by the test data of the clay; ξ_v = the trajectory of volumetric plastic strains.

(2) Unloading

When the clay experiences unloading, the constitutive stiffness tensor is assumed to be the elastic tensor, since laboratory tests on the clay showed that the unloading response was essentially elastic.

(3) Reloading

For the reloading, the corresponding incremental stress-strain relationship is given by

$$d\sigma_{ij} = C_{ijkl}^{RL} d\varepsilon_{kl} \tag{7}$$

where

$$C_{ijkl}^{RL} = C_{ijkl}^e - \frac{C_{ijnm}^e n_{nm}^R n_{op}^R C_{opkl}^e}{H^{RL} + n_{rs}^R C_{rstu}^e n_{tu}^R} \tag{8}$$

where n_{ij}^R = the unit normal tensor for the reference surface R; H^{RL} = the plastic modulus for reloading. For the Sabine Clay,

$$R = \left(\frac{J_{2D}}{P_a^2} \right) + \alpha_r \left(\frac{J_1}{P_a} \right)^n - \gamma \left(\frac{J_1}{P_a} \right)^2 \tag{9}$$

$$H^{RL} = H_{I_1}^{VL} + H_{I_2}^{VL} r_1 \left(1 - \frac{\alpha_{ps}}{\alpha_r} \right)^{r_2} \tag{10}$$

$$H_{I_1}^{VL} = \frac{-\dfrac{\partial R}{\partial \alpha_r}}{\left(\dfrac{\partial R}{\partial \sigma_{ij}} \dfrac{\partial R}{\partial \sigma_{ij}} \right)^{\frac{1}{2}}} \left(\frac{\alpha_r}{\alpha_{ps}} \right)^{\frac{n-1}{n-2}} \frac{\partial \alpha_{ps}}{\partial \xi_v} \frac{n_{kk}^F}{\sqrt{3}} \tag{11}$$

$$H_{I_2}^{VL} = - \frac{\alpha_{ps}^{-\frac{n-1}{n-2}} \gamma^{\frac{1}{n-2}}}{\sqrt{3}(n-2)} \frac{\partial \alpha_{ps}}{\partial \xi_v} \tag{12}$$

where r_1 and r_2 are the interpolation parameters for reloading and can be found from the test data for the clay. The value of α_r is obtained by setting $R = 0$ and substituting current stresses into the equation for R.

INCREMENTAL EQUATIONS OF MOTION FOR FOUNDATION

The derivation of the dynamic equilibrium equations for the foundation subsystem is based on the assumption that the dissipation of seismic energy through inelastic deformation tends to overshadow the dissipation of the energy through viscous damping (Anderson, 1989). Therefore, the velocity related damping terms in the dynamic equilibrium equations are neglected. The equations of motion derived with this assumption will yield conservative results for the response. The incremental equations of motion for the foundation at time $t + \Delta t$ can be expressed as follows:

$$\left(\frac{4M}{\Delta t^2} + {}^tK \right) \Delta V^{(k)} = {}^{t+\Delta t}R - {}^{t+\Delta t}F^{(k-1)} - M \left(\frac{4({}^{t+\Delta t}V^{(k-1)} - {}^tV)}{\Delta t^2} - \frac{4\,{}^t\dot{V}}{\Delta t} - {}^t\ddot{V} \right)$$

$$\tag{13}$$

where ${}^tK$ = tangential stiffness matrix of the foundation at time t; M = mass matrix of the foundation; Δt = time step; ${}^{t+\Delta t}R$ = the externally applied nodal dynamic loads, including both basement rock excitation and pile head forces; ${}^{t+\Delta t}F$ = nodal forces due to the element stresses; V, V', and V" = nodal dynamic displacement, velocity, and acceleration, respectively; k = iteration step; $\Delta V^{(k)}$ = the increment of V at the kth iteration. Therefore, the nodal dynamic displacements at the kth iteration of time step $t + \Delta t$ are ${}^{t+\Delta t}V^{(k)} = {}^{t+\Delta t}V^{(k-1)} + \Delta V^{(k)}$

COMPUTERIZATION OF PROPOSED METHODOLOGY

In order to implement the proposed three-dimensional nonlinear finite element subsystem methodology, a comprehensive and powerful computer program is essential. Since the proposed methodology is the first of its type, it was not possible to find commercial software which offers all of the required features as described in the following paragraph. Therefore, a comprehensive nonlinear Fortran program is developed in this study.

The basic features of the program are as follows. The program consists of more than seventy fundamental subroutines which can be grouped as several major functional modules such as Data Acquisition and Equation Profile Module, Seismic Analysis Module, Foundation Subsystem Module, Structure Subsystem Module, Eigenvalue Searching Module, and Equation Solution Module, etc. These modules can be modified or changed as desired. Therefore, the program is able to be updated with the progress of the research.

First, the Data Acquisition and Equation Profile Module reads the input data such as geometry, material, etc., and generates the finite element meshes and the initial property matrices for the foundation subsystem and the structure subsystem by calling the Foundation Subsystem Module, the Structure Subsystem Module, and the Eigenvalue Searching Module. Then, the Seismic Analysis Module reads the bedrock motion data and conducts the dynamic analysis by calling the Equation Solution Module, the Foundation Subsystem Module, and the Structure Subsystem Module.

For each time step, as convergence of the solution is reached, the Foundation Subsystem Module and the Structure Subsystem Module as well as the Eigenvalue Searching Module will be called by the Seismic Analysis Module to update the property matrices for next time step. The Seismic Analysis Module will terminate the running of the program when the entire response history is determined.

SOLUTION OF NONLINEAR DYNAMIC EQUATIONS

For seismic response analysis, the assumptions of large displacements and finite strains are desirable. However, considering the complexity of the proposed nonlinear soil model, which is the first of its type used in the soil-pile-structure interaction, and considering the analysis efficiency as well as the availability of nonlinear soil theories, a materially-nonlinear only formulation is selected for the study. The analysis based on the assumption will provide some insight into the contribution of soil nonlinearity to the dynamic soil-pile-structure interaction.

The basic problem in a general nonlinear analysis is to find the state of equilibrium of a body corresponding to the applied loads. The dynamic equilibrium conditions of a system of finite elements representing the body under consideration can be expressed as

$$^{t+\Delta t}P - {}^{t+\Delta t}F = 0 \tag{14}$$

where $^{t+\Delta t}P$ is a vector representing the externally applied nodal loads (including inertial and damping forces for dynamic problems) in the configuration at time $t+\Delta t$; and $^{t+\Delta t}F$ is a vector of nodal forces that correspond to the element stresses in this configuration. Since the nodal forces $^{t+\Delta t}F$ depend nonlinearly on the nodal displacements, it is necessary to iterate for the solution of Equation (14).

By means of modified Newton-Raphson iteration method (Bathe, 1982), the incremental dynamic equilibrium equation for the ith iteration at the time $t+\Delta t$ can be derived as follows:

$$(\ {}^{t}K + 4\frac{M}{\Delta t^2} + 2\frac{{}^{t}C}{\Delta t}) \Delta U^{(i)} = \ {}^{t+\Delta t}P - \ {}^{t+\Delta t}F^{(i-1)} - \ {}^{t}C[\frac{2}{\Delta t}(\ {}^{t+\Delta t}U^{(i-1)} - \ {}^{t}U) - \ {}^{t}\dot{U}] -$$

$$- M[\frac{4}{\Delta t^2}(\ {}^{t+\Delta t}U^{(i-1)} - \ {}^{t}U) - 4\frac{{}^{t}\dot{U}}{\Delta t} - \ {}^{t}\ddot{U}]$$

(15)

For the structure subsystem, the externally applied dynamic forces ${}^{t+\Delta t}P$ can be represented by the effective forces due to the foundation (pile head) motions as

$$ {}^{t+\Delta t}P_{eff} = - M_S R_S \ {}^{t+\Delta t}\ddot{U}_g \qquad (16)$$

where M_S is the mass matrix of the structure subsystem; R_S is the pseudostatic response influence coefficient matrix; and ${}^{t+\Delta t}\ddot{U}_g$ is the foundation (pile head) acceleration vector at the time $t+\Delta t$.

Therefore, the incremental dynamic equilibrium equation of the structure subsystem for the ith iteration at the time $t+\Delta t$ can be written as follows:

$$(\ {}^{t}K_S + 4\frac{M_S}{\Delta t^2} + 2\frac{{}^{t}C_S}{\Delta t}) \Delta U^{(i)} = - M_S R_S \ {}^{t+\Delta t}\ddot{U}_g - \ {}^{t+\Delta t}F^{(i-1)} - \ {}^{t}C_S[\frac{2}{\Delta t}(\ {}^{t+\Delta t}U^{(i-1)} - \ {}^{t}U) - \ {}^{t}\dot{U}] -$$

$$- M_S[\frac{4}{\Delta t^2}(\ {}^{t+\Delta t}U^{(i-1)} - \ {}^{t}U) - 4\frac{{}^{t}\dot{U}}{\Delta t} - \ {}^{t}\ddot{U}]$$

(17)

where the subscript S denotes the properties corresponding to the structure subsystem and U represents the nodal displacement vector of the structure subsystem.

For the foundation subsystem, the externally applied dynamic forces, ${}^{t+\Delta t}P$, include the effective forces due to bedrock excitation and the pile head forces transmitted from the structure subsystem, which can be expressed as

$$ {}^{t+\Delta t}P = -M_F R_F \ {}^{t+\Delta t}\ddot{V}_b + \ {}^{t}F_p \qquad (18)$$

where M_F is the mass matrix of the foundation subsystem; R_F is the pseudostatic response influence coefficient vector; ${}^{t+\Delta t}\ddot{V}_b$ is the bedrock acceleration at time $t+\Delta t$; and ${}^{t}F_p$ is the pile head force vector at time t.

Therefore, the incremental dynamic equilibrium equation of the foundation subsystem for the ith iteration at the time $t+\Delta t$ can be written as follows:

$$({}^{t}K_{F} + 4\frac{M_{F}}{\Delta t^{2}})\, \Delta V^{(i)} = -M_{F}R_{F}\,{}^{t+\Delta t}\ddot{V}_{b} + {}^{t}F_{p} - {}^{t+\Delta t}F^{(i-1)} -$$

$$- M_{F}[\,\frac{4}{\Delta t^{2}}(\,{}^{t+\Delta t}V^{(i-1)} - {}^{t}V) - 4\frac{{}^{t}\dot{V}}{\Delta t} - {}^{t}\ddot{V}]$$

(19)

noting that the velocity related damping terms have been neglected based on the preceding discussion. In Equation (19), the subscipts F denote the properties corresponding to the foundation subsystem, and V the nodal displacement vector of the foundation subsystem, respectively.

The pile head acceleration ${}^{t+\Delta t}\ddot{V}_{p}$ of the foundation subsystem is equivalent to the foundation acceleration ${}^{t+\Delta t}\ddot{U}_{g}$ of the structure subsystem. The pile head forces ${}^{t}F_{p}$ of the foundation subsystem are equivalent to the support forces ${}^{t}F_{g}$ of the structure subsystem.

CONVERGENCE CRITERIA

To terminate the iteration at an acceptable equilibrium condition, adequate convergence criteria should be checked at the end of each iteration (Bathe and Cimento, 1980). Considering the nonlinearity of the soil model used in the proposed methodology, three types of convergence criteria are used simultaneously to check the convergence of the iteration. These are the displacement criterion, the out-of-balance load criterion, and the internal energy criterion.

The displacement criterion can be written as

$$\frac{\| \Delta U^{(i)} \|_{2}}{\| {}^{t+\Delta t}U^{(i)} \|_{2}} \leq \epsilon_{D}$$

(20)

where the numerator on the left hand side is an Euclidean vector norm of $\Delta U^{(i)}$ and the denominator on the left hand side is an Euclidean vector norm of ${}^{t+\Delta t}U^{(i)}$; ϵ_{D} is a displacement convergence tolerance.

The out-of-balance load criterion may be expressed as

$$\frac{\| {}^{t+\Delta t}R - {}^{t+\Delta t}F^{(i)} \|_{2}}{\| {}^{t+\Delta t}R - {}^{t}F \|_{2}^{\max}} \leq \epsilon_{F}$$

(21)

where the numerator on the left hand side is an Euclidean vector norm of ${}^{t+\Delta t}R - {}^{t+\Delta t}F^{(i)}$ and the denominator on the left hand side is a maximum Euclidean vector norm of ${}^{t+\Delta t}R - {}^{t}F$ which is always calculated during the solution; ϵ_{F} is an out-of-balance load tolerance.

Since the incremental displacements and out-of-balance loads may change independently, an internal energy criterion may be necessary to provide some indication of when both the displacements and the forces are near their equilibrium values. The internal energy criterion may be written as

$$\frac{\Delta U^{(i)T}(^{t+\Delta t}R \ -^{t+\Delta t}F^{(i \ - \ 1)})}{\Delta U^{(1)T}(^{t+\Delta t}R \ -^{t}F)} \ \le \ \epsilon_E \tag{22}$$

where the superscript T of ΔU denotes the transposition of the vector ΔU; ϵ_E is an energy tolerance.

DATA STORAGE AND EQUATION SOLVER

A major concern in the three-dimensional finite element analysis is the availability of computer storage space. To deal with this problem efficiently, the proposed methodology employs the subsystem concept to reduce the space needed for property matrices. In addition, some effective computer storage allocation and matrices storage schemes are used to further reduced the storage space needed.

A single array is set up in the data acquition module and partitioned to store all the data arrays and the global arrays such as nodal coordinates, nodal displacements, nodal velocities, nodal accelerations, masses, stiffnesses, etc. Each array is variably dimensioned to the exact size required for each program by using a set of pointers established in the control program. In this way, no space is wasted in data storage and a maximum amount of space is reserved to store the global arrays.

To save the space used in the storage of the matrices, a Profile or Skyline storage scheme (Zienkiewicz and Taylor, 1989; Bathe, 1982) is adopted in the program. Since the associative model of the δ^*-vertion of the HiSS approach is used to describe the constitutive law of the soil in the study, the resulting constitutive tensor C^*_{ijkl} will be symmetric and, in turn, the stiffness matrices in the finite element procedures will also be symmetric. Therefore, by using the profile storage scheme, only the items under the non-zero profile (skyline) and above the diagonal (including diagonals) of the stiffness (or mass and damping) matrices are stored in the analysis procedure. This storage scheme has definite advantages over the most of other storage methods. It always requires less storage and the storage requirements are not severely affected by a few very long columns of the matrices.

Because the program is expected to perform nonlinear analysis of the seismic response of the interactive soil-pile-structure systems and the nonlinear solution is usually approximated by a series of iteration of the solutions of the linear incremental equations, the efficiency of the solver for the simultaneous equations is of vital importance. Therefore, an active column profile (skyline) solution algorithm is employed in the Equation Solution Module to solve the equations efficiently. This solution algorithm is based upon the Crout method of Gauss elimination. The method consists of a factorization of the stiffness matrix (or a equivalent stiffness matrix in the iteration solution of nonlinear dynamic equations) into the product of a lower triangular matrix and an upper triangular matrix.

The use of this method with an active column profile storage scheme leads to a very compact program, and it is very easy to use vector dot product routines to effect the triangular decomposition and forward reduction. This computational advantage is very important to modern computers which are vector oriented. Another attractive advantage of the active column profile solution scheme is that it allows for the inclusion of a re-solve capability (i.e., new load increments for the iteration loops) without any significant additional programming effort. Use of the re-solve capability can substantially reduce costs for analyzing subsequent load increments in the iteration solution procedure for nonlinear dynamic analysis.

IMPLEMENTATION OF THE FOUNDATION MODEL

The foundation subsystem consists of piles and surrounding soil which are idealized as an assemblage of eight-node isoparametric hexahedral elements. The major task in the implementation of the foundation subsystem model is the numerical modeling of the nonlinear soil properties.

For the proposed soil model, the HiSS approach is used in the numerical procedure to determine the updated constitutive stiffness tensor. The principles of plasticity theory (Chen and Baladi, 1985; Chen and Han, 1988) are followed in the implementation of the HiSS model of the soil. At the beginning of the each time step, the constitutive stiffness tensor, C^*_{ijkl} (C^{VL}_{ijkl} or C^{RL}_{ijkl}), is updated based on the current strain-stress situation of the soil. Since the C^*_{ijkl} will take different formulations for different loading conditions, the loading condition at a Gauss point of the soil element should be determined first.

There are three possible loading conditions, i.e. virgin loading, reloading, and unloading. When a stress point is on the yield surface, the loading conditions may be virgin loading or unloading; when a stress point is inside the yield surface, reloading, reloading followed by the virgin loading, or unloading may occur.

The location of a stress point in the stress space is determined by checking the current stress tensor, σ_{ij}, with the yield function $F(\sigma_{ij}, \alpha_{ps})$, Equation (5). The hardening function, α_{ps}, may take the value from the last yield surface. If $F(\sigma_{ij}, \alpha_{ps}) = 0$, the stress point is on the yield surface; if $F(\sigma_{ij}, \alpha_{ps}) < 0$, the stress point is inside the yield surface.

When a stress point is on the yield surface, the direction of the stress increment, $d\sigma_{ij}$, will be checked to determine whether the strain increment, $d\epsilon_{kl}$, will cause virgin loading or unloading. The exact direction of $d\sigma_{ij}$ for a given $d\epsilon_{kl}$ is not known in advance. Therefore, an approximate direction is evaluted by using an elastic predictor stress increment, $d\sigma^e_{ij}$. It may be obtained by the following equation (Wathugala and Desai, 1990):

$$d\sigma^e_{ij} = C^e_{ijkl} d\varepsilon_{kl} \tag{23}$$

Equation (23) is the incremental form of the first order of the Cauchy elastic law, which is the same as the generalized Hooke's law (Desai and Siriwardane, 1984). The elastic constitutive stiffness tensor C^e_{ijkl} can be expressed as (Chen and Han, 1988)

$$C^e_{ijkl} = \frac{E}{2(1+\upsilon)} \left[\frac{2\upsilon}{1-2\upsilon} \delta_{ij} \delta_{kl} + \delta_{ik} \delta_{jl} + \delta_{il} \delta_{jk} \right] \qquad (24)$$

where E = the Young's Modulus; υ = the Poisson's Ratio; δ_{ij} = the Kronecker delta

$$\delta_{ij} = \begin{cases} 1 & if \quad i = j \\ 0 & if \quad i \neq j \end{cases} \qquad (25)$$

Then the direction of the stress increment will be determined by checking the values of $d\sigma^e_{ij} n^F_{ij}$, $F(\sigma^e_{ij}, \alpha_{ps})$, and α^e_r. α^e_r is the hardening function value for the reference surface, R, passing through the elastic predictor. n^F_{ij} is the unit normal to the last yield surface, F.

$$n^F_{ij} = \frac{\dfrac{\partial F}{\partial \sigma_{ij}}}{\left(\dfrac{\partial F}{\partial \sigma_{mn}} \dfrac{\partial F}{\partial \sigma_{mn}} \right)^{\frac{1}{2}}} \qquad (26)$$

Generally, if $d\sigma^e_{ij} n^F_{ij} > 0$, $F(\sigma^e_{ij}, \alpha_{ps}) > 0$, and $\alpha^e_r < \alpha_{ps}$, the stress point is under virgin loading; if $d\sigma^e_{ij} n^F_{ij} = 0$, $F(\sigma^e_{ij}, \alpha_{ps}) > 0$, and $\alpha^e_r < \alpha_{ps}$, the stress point is under neutral loading; if $d\sigma^e_{ij} n^F_{ij} < 0$, $F(\sigma^e_{ij}, \alpha_{ps}) < 0$, and $\alpha^e_r > \alpha_{ps}$, the stress point is under unloading.

To avoid the numerical difficulty encountered in situations where the large strain increment passes through different types of loading conditions, a subincrement technique (Desai, Muqtadir, and Scheele, 1986; Wathugala and Desai, 1990) is used to reduce the strain increment. The stress path is evaluated corresponding to each subincrement of the strain.

When a stress point is inside the yield surface, the possible loading conditions are assumed to be reloading, neutral loading, and unloading, provided that the subincrement technique is used, and the possible virgin loading following the reloading will be treated as the virgin loading occured on the yield surface.

If $d\sigma^e_{ij} n^F_{ij} > 0$, $F(\sigma^e_{ij}, \alpha_{ps}) < 0$, and $\alpha_r > \alpha^e_r > 0$, the stress point is under reloading; if $d\sigma^e_{ij} n^F_{ij} = 0$, $F(\sigma^e_{ij}, \alpha_{ps}) < 0$, and $\alpha_r > \alpha^e_r > 0$, the stress point is under neutral loading; if $d\sigma^e_{ij} n^F_{ij} < 0$, $F(\sigma^e_{ij}, \alpha_{ps}) < 0$, and $\alpha^e_r > \alpha_r > 0$, the stress point is under unloading.

After the loading conditions are determined, the stress state for nonvirgin loading conditions may be obtained using the corresponding constitutive tensors. For virgin loading cases, an elastic-predictor-plastic-corrector approach is used with an iteration procedure to determine the new yield surface.

The constitutive stiffness tensor at the point will then be updated for the corresponding stress state. For a stress point located on the new yield surface, the new hardening function, α_{ps}, is first determined for the current trajectory of volumetric plastic strains, ξ_v. ξ_v can be calculated from the increment of ξ_v, which is defined as

$$d\xi_v = \begin{cases} \dfrac{|d\epsilon^p_v|}{\sqrt{3}} & \text{for } d\epsilon^p_v > 0 \\ 0 & \text{for } d\epsilon^p_v \leq 0 \end{cases} \tag{27}$$

where $d\epsilon^p_v$ is the incremental volumetric plastic strain due to virgin loadings.

For the associative model of the HiSS approach, the potential function Q is assumed to be same as the yield function F. Therefore, the incremental volumetric plastic strain $d\epsilon^p_v$ can be calculated from the flow rule of the plasticity as

$$d\epsilon^p_{kl} = \lambda\, n^F_{kl} \tag{28}$$

where λ is a scalar of proportionality, which can be obtained approximately from the iteration procedure for the virgin loading cases; n^F_{kl} is the tensor of the unit normals to the current yield surface, which can be found from Equation (26).

Then the virgin plastic modulus H^{VL} can be found from the Equation (4) and the constitutive stiffness tensor for the virgin loading can be calculated by Equation (3). All of the data related to the current yield surface will be saved for the subsequent time steps until a new yield surface is encountered again. If the loading condition for the next (current) time step is a virgin loading, the updated virgin constitutive stiffness tensor will be used. And if unloading occurs, the unloading stiffness tensor will be used.

For a stress point inside the last yield surface, the current hardening function α_r is determined from the current reference surface R, which is the function of the current stress state. The data related to the last yield surface are retrieved from the last yield surface data saved in the program. Then the reloading constitutive stiffness tensor is computed from Equations (12), (11), (10), and (8). If the next (current) time step is reloading again, the current reloading constitutive stiffness tensor will be used. And if the reloading should cause a new virgin loading, the last virgin loading constitutive stiffness tensor will work for the case. When an unloading occurs, the unloading constitutive stiffness tensor will be used.

SUMMARY

A three-dimensional nonlinear methodology for the seismic Soil-Pile-Structure-Interaction (S-P-S-I) has been developed. To implement the proposed methodology, a comprehensive and powerful computer program has been created. The major aspects of the numerical modelling procedures for the program are presented in this paper. The computerized methodology can be utilized to investigate the seismic response of interactive Soil-Pile-Structure systems.

APPENDIX. REFERENCES

Anderson, J. C. (1989). "Dynamic Response of Buildings." *The Seismic Design Handbook* (Ed. by F. Naeim), Van Nostrand Reinhold, New York, NY, 113-118.

Bathe, K.J. and Cimento, A.P. (1980). "Some Practical Procedures for the Solution of Nonlinear Finite Element Equations." *J. Computer Methods in Applied Mechanics and Engineering*, vol. 22, pp. 59-85.

Bathe, K. J. (1982). *Finite Element Procedures in Engineering Analysis*, Prentice-Hall, Inc., Englewood Cliffs, NJ.

Cai, Y.X. (1995). *Three-Dimensional Nonlinear Analysis of Seismic Soil-Pile-Structure Interaction and Its Application*, Doctoral Dissertation, Washington University, St. Louis, USA.

Cai, Y.X., Gould, P.L., and Desai, C.S. (1995). "Investigation of 3-D Nonlinear Seismic Performance of Pile-Supported Structures." ASCE Annual Convention, San Diego, CA, USA.

Chen, W. F., and Baladi, G. Y. (1985). *Soil Plasticity - Theory and Implementation*, Elsevier, Inc., New York.

Chen, W. F., and Han, D. J. (1988). *Plasticity for Structural Engineers*, Springer-Verlag New York Inc.

Desai, C. S., and Siriwardane, H. J. (1984). *Constitutive Laws for Engineering Materials*, Prentice-Hall Inc., Englewood Cliffs, New Jersey.

Desai, C. S., Muqtadir, A. and Scheele, F., (1986). "Interaction Analysis of Anchor Soil Systems." *J. Geotech. Engrg.*, ASCE, vol. 112, GT5.

Desai, C.S., and Wathugala, G.W. (1993), "Constitutive Model for Cyclic Behavior of Clays. II: Applications", *J. of Geotech. Engrg.*, ASCE, Vol. 119(4), 730-748.

Wathugala, G. W. and Desai, C. S. (1990). "Dynamic Analysis of Nonlinear Porous Media with Anisotropic Hardening Constitutive Model and Application to Field Tests on Piles in Saturated Clays." Report to NSF, CEEM Dept., Univ. of Arizona, Tucson, Az.

Wathugala, G.W., and Desai, C.S. (1993), "Constitutive Model for Cyclic Behavior of Clays. I:Theory", *J. of Geotech. Engrg.*, ASCE, Vol. 119(4), 714-729.

Zienkiewicz, O.C., and Taylor, R. (1989). *The Finite Element Method*, 4th ed., McGraw-Hill, New York.

Seismic Performance of Integral Abutment Bridges

Jay Shen[1] and Manuel Lopez[2]

Abstract

The seismic behavior of an integral abutment bridge is studied by taking into account the effects of the soil, piling and abutments. The overall seismic behavior of the structure is then compared to the seismic behavior of the same structure without considering the soil and piling effects.

Introduction

Expansion joints on bridges prevent damage caused by thermal expansion and contraction of the superstructure from temperature variations. However, joints increase the initial and maintenance costs. Thus, integral abutment bridges provide an attractive and economic design alternative since expansion joints are eliminated from the structure.

In spite of the fact that integral abutment bridges have acceptable performance under seismic loads, there is no model to predict or even to study their dynamic behavior analytically. The design of an integral bridge is often based on intuitive practice rather than systematic investigation. A realistic design approach is difficult to obtain since there are uncertainties in both the piling and the abutment behavior under lateral loads. In addition these are problems in the soil-structure interaction field.

A complete non-linear 3-dimensional model, including the bridge, footing and surrounded soil, was developed. The finite-element program ABAQUS was used to

[1] Ph D. SE and PE, Assistant Professor, Department of Civil and Architectural Engineering, Illinois Institute of Technology, 3201 S. Dearborn Street, Chicago, IL 60616. Tel: (312) 567-5860; Fax: (312) 567-3519; Email:jshen@charlie.cns.iit.edu.
[2] Graduate Research Assistant, Department of Civil and Architectural Engineering, Illinois Institute of Technology, 3201 S. Dearborn Street, Chicago, IL 60616.

perform the nonlinear static and dynamic analyses. The characteristics of the model are explained in the following section.

Modeling

An actual integral abutment bridge was chosen for the analysis. The structure is a 2 span bridge with a total length of 144.2 m. Each span measures 71.6 m. The structure consists of the following elements:

- Three steel plate girders each 144.2 m long and 1.83 m deep which are spaced 3.2 m apart.
- The steel plate girders support a 21.6 cm composite concrete deck slab.
- Each girder's top flange measures 45.7 cm wide by 2.86 cm thick and the bottom flange is 45.7 cm wide by 8.3 cm thick.
- The web of the plate girders has a depth of 1.83 m and is 1.27 cm thick.
- The vertical cross frames are of inverted chevron type with and area of 33.9 squared centimeters placed at interval of 7.16 m apart.
- The abutments of the bridge are supported by steel piles.
- The piles from the southern abutment are 12.57 m long while the piles from the northern abutment are 15.74 m long. The section of the piles is HP12X53.
- The type of the soil where the piles were driven is predominantly clay.
- Granular material was placed at the abutment backfills.
- The structure is supported at its middle by a concrete pier. This pier has a uniform section of 1.067 m. The joint between the superstructure and the pier is considered as pinned.
- The bottom of the pier is based on bedrock.
- The pier is supported by two piles of 2.32 m long.

Drawings from the bridge are shown in Figures 1 to 6.

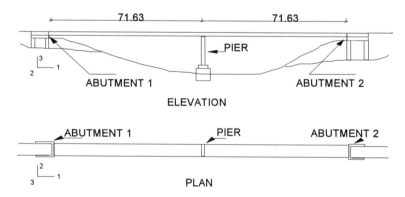

Figure 1. Plan and Elevation Views of the Bridge

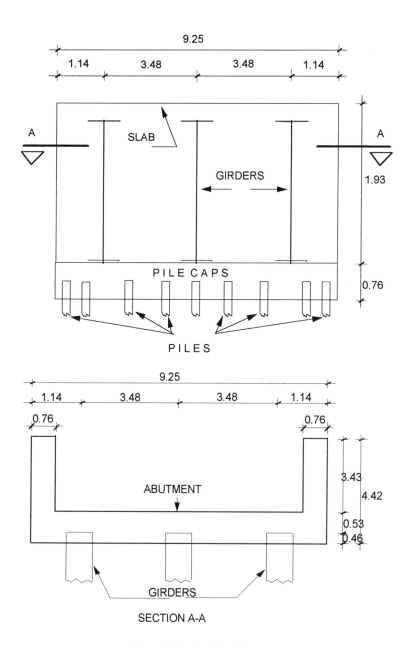

Figure 2. Detail of the Abutment

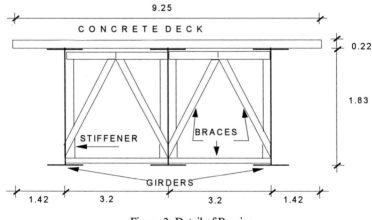

Figure 3. Detail of Bracing

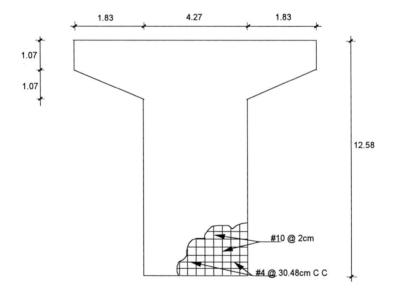

Figure 4. Detail of the Pier
Note: All the dimensions are in meters except the reinforcement detailing.

The composite deck is modeled using shell elements. The material assigned to these shells is concrete. See Figure 5. Shell elements are utilized to model the steel girders. The properties assigned to the steel are shown in Figure 5. In regard to the brace modeling, a two component truss element is conceived to capture the differences between the tension behavior and the compression behavior. Although the tension component material has similar properties than those possessed by the girder model material, the compression component present values to account for the buckling of the brace.

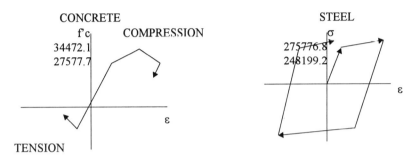

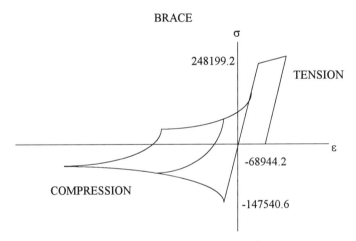

Figure 5. Material and Brace Models
Note: All stresses are in KPa

Abutments and pile caps are represented by shell elements. The material considered for these members is concrete. The characteristics of this concrete are similar to those of the modeled slab concrete. Similarly, the pier has been represented by using shell elements. The material used is concrete; however, its first yielding stress is considered to happen at 13788.8 KPa. The material, then, is considered to harden to a stress of 15167.7 KPa. The considered thickness of the slab is 0.762 m.

This structure is supported at its abutments by piles. The piles are modeled using beam elements. The material is steel with similar properties to the steel used in the modeling of the beams. The pile sections are HP12X53. Although there are eleven piles per abutment, nine piles are considered in the model. Each extreme pile, at the abutments, accounts for the two piles that are located in the lateral walls of the abutment.

In this model, the soil surrounding the piles and the soil at the abutment backfill have modeled. In the process, truss elements have been utilized to capture the soil behavior. The trusses account for:

- The relationship between the lateral soil pressure and the corresponding lateral pile displacement (p-y curves). See Figure 6.
- The relationship between the skin friction and the relative vertical displacement between the pile and the soil (t-z curves). See Figure 7.
- The relationship between the bearing stress at the pile tip and the pile tip settlement (q-z curves). See Figure 8.

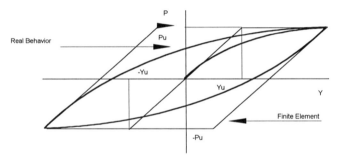

Figure 6. Typical p-y Cyclic curve of the Model and Actual Behavior.

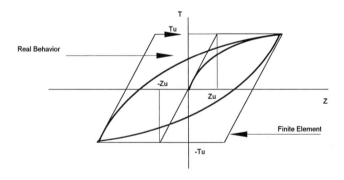

Figure 7. Typical t-z Cyclic Curve of the Model and the Actual Behavior.

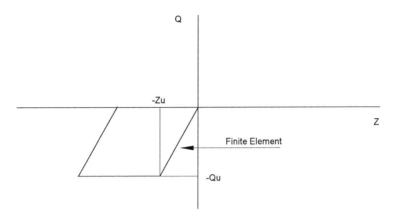

Figure 8. Typical q-z Cyclic Curve of the model

In addition to near 1730 trusses used in the soil-pile modeling, each abutment wall has 5 additional trusses. These elements account for the backfill effects. There are two trusses that simulate the effects of the soil in the lateral walls and other three that idealize the effects of the backfill soil on the central wall. The lateral trusses work both in tension and in compression; however, the three central trusses have compression component only.

Besides the modeling of the bridge footing and surrounded soil is accomplished the self weight of the structure is considered in all analysis. This fact produces more realistic representation of the actual structure. Figure 9 Shows the final 3-dimensional model. Total weight of the structure is about 13200 KN.

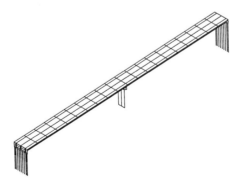

Figure 9. Complete Nonlinear 3-dimensional Model

Dynamic Seismic Analysis

The response of the bridge is studied under different excitations.   Two representative ground motions, the north-south component of both El Centro and Northridge/Newhall earthquakes, are used in this section. Figure 9 gives the plots of the normalized response spectra with two percent damping of the two ground motions. It can be observed that the Newhall motion excites the structures with large range of frequencies from 0.3 seconds up to 1.4 seconds while El Centro motion excites structures with frequencies from 0.1 seconds to 1 second.

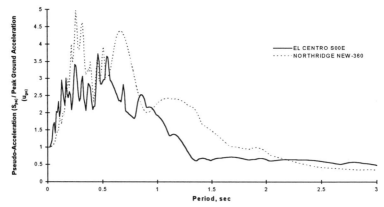

Figure 9. Normalized Response Spectra for El Centro and Newhall Earthquakes

In order to study the behavior of the bridge subjected to different severity of excitations, the intensities of the ground motions are increased gradually.   For instance El Centro, with peak ground acceleration of 0.35g, is factored by 1.0, 1.5 and 2.0.   Similarly, Newhall with peak ground acceleration of 0.59g is factored by 1.0, 1.5 and 2.0.   These six excitations are applied in the transverse (Y) direction of the model.   A time period of 20 seconds is considered in each analysis.   Figures 10 and 11 show the time history for both ground motions.

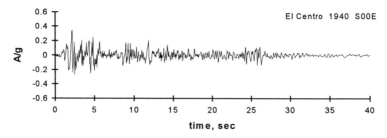

Figure 10. Time History of El Centro North-South Component

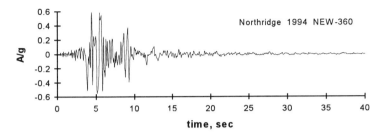

Figure 11. Time History of the Newhall North-South Component

Results

During the process of analysis there has been an interesting fact which is the large displacements experienced by the concrete deck above the pier. In order to ponder the effect of the soil-structure interaction in the structure a new model of the bridge is used. In this representation, the soil and piling have not been considered and the structure is supposed to have fixed base. Figure 12 shows a comparative plot between the response of the bridge considering the soil-structure-interaction versus response of the bridge considering the structure having fixed supports. Each point in the figure represents the peak base shear versus maximum displacement obtained by the time history analysis.

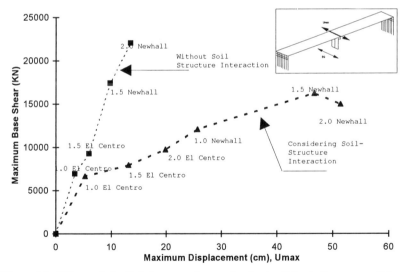

Figure 12. Effects of the Soil-Structure Interaction in the Seismic Response of the Bridge

Displacements from the soil-structure-interaction plot are considerably larger than the displacements from the plot where no soil-structure-interaction is considered. This variation can be explained from the torsion that the abutment experienced when the soil modeling is considered. The effect can be compared to have a straight bar supported by springs at its both extremes. If a unit force is applied to its center, the displacement at that point will be the result of the deformation due to the structure itself plus the rotation allowed by the springs at its extremes. In order to clarify this concept an example is presented.

Considered the maximum longitudinal displacements of the nodes at the extremes of an abutment for the 1.5 Newhall excitation. See figure 13.

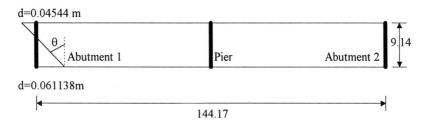

Figure 13. Displacements at the Extremes of the Abutment 1

The angle θ is equal to 1.17%. The deformation due to a unit load applied at the middle of the structure can be calculated as follows:

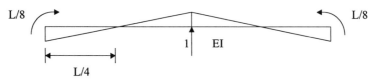

Figure 14. Diagram of the Loading

From the figure above, the displacement at the middle of the bar can be found by the formula:

$(1.\Delta)=\Sigma\int (EI/L)+2(L/8)\theta$. Then: $\Delta=0+(144.17/4)(1.17\%)=0.422$m

The actual displacement of the slab is 0.4663m, value that is close to 0.422m.

The large deformations at the slab impose demands of strength over the elements close to and above the pier. In consequence, these members are the most susceptible elements to suffer damages in a seismic excitation.

Figures 15 to 18 show the behavior of selected brace elements, subjected to a seismic excitation equal to 1.5 Newhall.

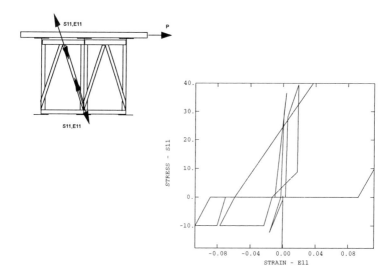

Figure 15. Behavior of a brace located above the pier

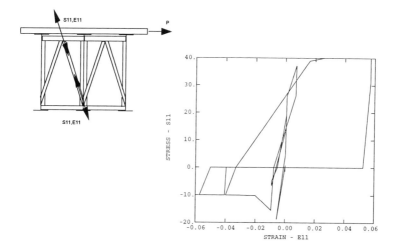

Figure 16. Behavior of a brace located next to the abutment

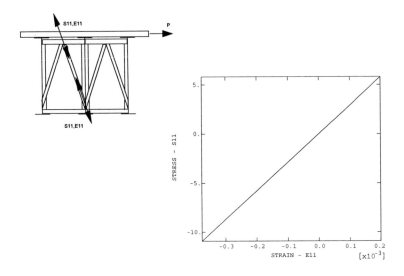

Figure 17. Behavior of a brace located between the pier and abutment 1

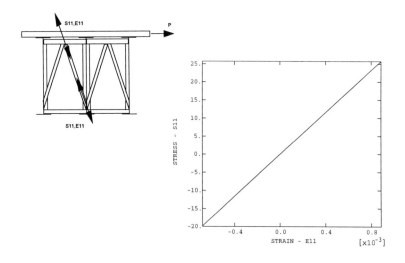

Figure 18.   Behavior of a brace located next to the abutment 1
Stresses in Ksi. 1Ksi = 6895 KPa.

Similarly, the rotation of the abutments produces demand in the pile elements that are located close to the abutment, at the top of piles. These piles can reach their plastic capacity in some seismic excitations due to the bending moments produced at the abutments. Figures 19 to 21 show the behavior of some critical elements located at top of piles. The seismic excitation used is 1.5 Newhall.

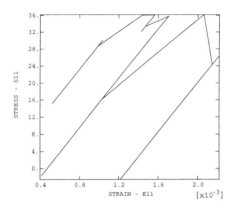

Figure 19. Behavior of element located at the extreme pile

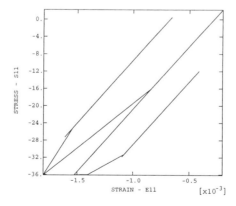

Figure 20. Behavior of element located in the pile next to the extreme pile

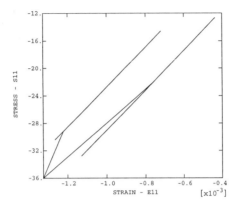

Figure 21. Behavior of element located in the third pile from the extreme pile
Stresses in Ksi. 1Ksi = 6895 KPa.

Conclusions

The seismic dynamic analysis showed that the brace elements above the pier
are the first one to reach their plastic capacity. Then, the nonlinear trend is extended
to the adjacent braces. The braces located at the midspan, between the pier and the
abutments, were the least stressed of the studied brace elements.

From the dynamic analysis was also detected that the all of the studied pile
elements reached their plastic behavior prematurely. The fact that piles have high
stresses produced by the self weight of the bridge could have accelerated these
failures.

In general, seismic performance of the integral bridges is satisfactory under
strong earthquakes.

Soil-Pile-Structure interaction has to be considered to access seismic behavior
of the integral bridge. Damage to piles may create functional problem under service
load after a major earthquake.

References

Abendroth, R.E., and Greimann, L.F., "Rational Design Approach for Integral Abutment Bridge Piles," Transportation Research Record, No. 1223, pp. 12-23, 1989.

Abendroth, R.E., and Greimann, L.F., "Abutment Pile Design for Jointless Bridges," Journal of Structural Engineering, Vol. 115, No. 11, pp. 2914-2929, Nov. 1989.

Burke, M.P. Jr., "Integral Bridges: Attributes and Limitations," Transportation Research Record, No. 1393, pp. 1-8, 1993.

Dagher, H.J., Elgaaly, M., and Kankam, J., "Analytical Investigation of Slab Bridges with Integral Wall Abutments," Transportation Research Record, No. 1319, pp. 115-125, 1991.

Greimann, L.F., Abendroth, R.E., Johnson, D.E., and Ebner, P.B., Pile Design and Test for Integral Abutment Bridges. Final Report, Iowa DOT Project HR-273, Iowa Department of Transportation, December 1987.

Greimann, L.F., and Wolde-Tinsae, A.M., "Design Model for Piles in Jointless Bridges," Journal of Structural Engineering, Vol. 114, No. 6, pp. 1354-1371, June 1988.

Greimann, L.F., Yang, P.S., and Wolde-Tinsae, A.M., "Nonlinear Analysis of Integral Abutment Bridges," Journal of Structural Engineering, Vol. 112, No. 10, pp. 2263-2280, Oct. 1986.

Hibbitt, Karlsson, and Sorensen, ABAQUS. User's Manual Volume I & II, Hibbitt, Karlsson & Sorensen, Inc., Version 5.4, 1994.

Hibbitt, Karlsson, and Sorensen, ABAQUS. Post Manual, Hibbitt, Karlsson & Sorensen, Inc., Version 5.4, 1994.

Lam, I.P., and Martin, Geoffrey R., Seismic Design of Highway Bridge Foundations. Volume II, Report FHWA/RD-86/102, FHWA Research Development Technology, 1986.

Lam, I.P., and Martin, Geoffrey R., Seismic Design of Highway Bridge Foundations. Volume III, Report FHWA/RD-86/102, FHWA Research Development Technology, 1986.

Wasserman, E.P., and Walker, J.H., "Integral Abutment for Steel Bridges," Highways Structures Design Handbook, Vol. II, Chap. 5, pp. 1-20, Oct. 1996.

SUBJECT INDEX

Page number refers to the first page of paper

AUTHOR INDEX

Page number refers to the first page of paper